John Roach

The American Carrying Trade

A plain talk to our public men and people who desire the revival of our

ocean carrying trade

John Roach

The American Carrying Trade
A plain talk to our public men and people who desire the revival of our ocean carrying trade

ISBN/EAN: 9783337319717

Printed in Europe, USA, Canada, Australia, Japan

Cover: Foto ©berggeist007 / pixelio.de

More available books at **www.hansebooks.com**

THE AMERICAN

CARRYING TRADE.

*A PLAIN TALK TO OUR PUBLIC MEN AND PEOPLE WHO
DESIRE THE REVIVAL OF OUR OCEAN CARRYING
TRADE, AND THE STEADY DEVELOPMENT OF
THE RESOURCES AND INDUSTRIES OF
THE UNITED STATES; AND WHO RECOGNIZE THE NEED
OF OPENING UP NEW MARKETS TO KEEP PACE
WITH THE NATION'S GROWTH.*

NEW YORK:

H. B. GROSE AND COMPANY.

1880.

PREFACE.

In presenting this monograph upon the American carrying trade on the ocean, the criticism may be forestalled that too much is said about the emergency of war and the policy of England. It may here be offered in apology, that it was through the emergency of war we lost that trade ; and that it was England who took it from us. It is, moreover, in the light of the policy that made England the first carrying nation of the world, and enabled her to seize every opportunity to extend her ocean empire, that we may most profitably study the condition of our own carrying trade, and consider what remedy exists for its present deplorable depression.

Aside from this, there is no other foreign nation to-day so closely watching our legislation, so deeply interested in it, or using such means to influence it, as England is,—the proof of her interest and concern disclosing itself abundantly through her press and public utterances. It will be well for this country if our own people are equally concerned and watchful to see that our legislation is in our own interest and for our own development and growth as a nation.

New York, *February,* 1880.

THE AMERICAN CARRYING TRADE.

I.

THE SUBJECT OUTLINED.

For fifteen years the depressed condition of our carrying trade on the ocean has been the subject of discussion in Congress, and through the debates in Committee and one means or other of delay much valuable time has been lost. But the time has come when, as was recently said by a United States Senator in a public speech in New York city, we can no longer afford to sit quiet and see our vast surplus products carried to the world's markets by the ships of our great commercial rival; when we must adopt a policy that will regain for us our lost carrying trade, and must boldly challenge our great competitor for the mastery of the sea.

At this juncture I beg leave to present for your notice some considerations touching upon this important question.

In 1879 our exports were 11,149,160 tons of surplus products; imports, 3,782,530 tons. This grand total of 14,931,690 tons was valued at about $1,550,000,000; of this $1,133,000,000 was carried in foreign ships, and only $417,000,000, or about one-third, in American ships. On the safe and speedy delivery in the markets of the world of these surplus products, and on the cash returns for them, rests the financial stability of our government. Block up for a year the road to these markets, and can there be any doubt that the resumption of specie payments would fail? In such event who can measure the disastrous effects that would follow? If the right of way were from any cause denied to us, what amount of life and treasure would not the nation sacrifice to regain it? Is it right, then, is it national, to trust so vast an interest to the hands of foreign carriers?

With eight millions of men in arms in Europe to-day, with immense navies ready to plunge into war at a day's notice, the foreign means of carrying our products may be cut off at any moment. War would at least advance the rates of freight enormously, while none of the extra profits would come to us, but would be drained from us. It was Washington's farewell advice that this nation should keep as free as possible from European complications. At the same time we should be prepared to profit by them, while being in no wise involved in them.

Again in 1877, our exports were 7,721,700 tons, our imports 3,593,804, valued at $1,173,000,000. Of this, foreign ships carried $858,000,000, American ships only $315,000,000. As against this showing, in 1851, of a total ocean trade of $433,000,000, our own ships carried $316,000,000, foreign ships $117,000,000. Thus in thirty years our position on the ocean has been exactly reversed. While our foreign trade has increased from less than $500,000,000 in 1851 to more than $1,100,000,000 in 1879, our carrying trade, so far from keeping pace with the general progress of the country, was actually smaller in 1877 than in 1851, and foreign ships were carrying over seven times as much of our products as they did in 1851, and gathering on the sea the golden results of our growth and prosperity on the land.

For fifteen years past not less than $60,000,000 a year have been paid by this country to foreign ship owners for carrying our mails, passengers, and freights, making a grand total of something like one thousand million of dollars. The sum now annually paid is estimated at from $60,000,000 to $75,000,000. These vast sums paid to foreign capitalists constitute a disastrous drain upon our financial resources. This drain had much to do with postponing the date of specie resumption, and every year of that postponement cost the country millions of dollars. I would ask the property owners, merchants, and mechanics of New York, where the largest part, at least three-fourths, of this enormous amount is paid out, what would be the benefit to them if the crews belonged in New York, if pay day came there, if the repairs to the fleet were made there, instead of in Liverpool as now ; and if, thus, a large part of this $60,000,000 or $70,000,000 was distributed in New York among our own people? And yet, strangely enough, the representatives of New York in Congress, and a large portion of the press, are blind to this great interest of their city in the matter, and advocate their rival's policy.

In 1857 our capitalists were expending some $25,000,000 a year in new ships, and a large sum in repairs. Now the amount annually spent in building ships is only about $11,000,000. In this connection, it should be remembered that 90 per cent. of the cost of a ship's construction goes direct to labor, and thus passes into the general circulation of the country.

These plain statements speak for themselves. That our position on the ocean to-day is inferior to what it was twenty-five years ago is a matter that may well command the study of our wisest statesmen. No other nation that was once a maritime power ever suffered such a decadence of its marine without either ruin or the loss of its importance among the peoples of the world. Our retrogression appears but the stranger when you consider that in natural resources for ship-building we are unequaled, and in skilled labor not surpassed by any country.

This shipping question is to-day the most important question before our people. For years it has been thrust aside by the pressing problems of politics and finance growing out of our civil war. Those problems are now nearly disposed of. Many other branches of industry which suffered by the war have been attended to, though vastly inferior in importance to this. The interest and attention of the country have been absorbed in the development of our exhaustless internal wealth, to the shameful neglect of this, one of the three great branches of national power and prosperity.

To the fact of this neglect and its consequences our people are now awakening. The revival of our carrying trade becomes, therefore, a vital issue, which no party can afford to treat with indifference or injustice. In this issue the whole people are concerned, for in it our national finances and industries, and our position among the nations, are alike involved. The whole people irrespective of party should meet it, therefore, fairly and without prejudice.

The ocean carrying trade has been for centuries the prize for which the leading nations of the world have contested. It has always been, and must be, the source of enormous wealth to the nation controlling it. By wise legislation and jealous fostering of the shipping interest England has made herself to-day practically the world's carrier on the seas. Her revenue in ocean freights and passage money and mail service is over $200,000,000 a year, and to earn this vast sum she drains not her soil, nor in the least impoverishes it by parting with any of its products. No matter what the financial condition of the country from which England collects this freight money, it must come to her in gold. Her policy, by which she controls this enormous business, has made her the banking house of the world. The gold which she thus draws from other nations, England often loans back to them at a large advance, taking therefor their bonds below par, but requiring both principal and interest to be paid in gold, and thus making out of them double profit. We have had some experience of this in the last twenty years.

Few men of the present generation know the struggles of our statesmen and people in the early days of the Republic to secure for her the proud and profitable place she held on the ocean for the first half century of our national existence. The policy they pursued achieved results that made the world wonder. It was reserved for our statesmen and people to-day to find the carrying trade wrested from us, and our flag almost driven from the seas. Who will find the remedy?

I earnestly ask your consideration of the following outline of the history of our rise and decline as a carrying nation, with some facts regarding the causes of that decline, and the true character of the movement now and of late in progress toward the repeal of our Navigation Laws. From these facts I hope to show you the falseness of proclaiming the remedy to lie in free ships, and the extent to which the Navigation Laws have aided and are now aiding us in our endeavors to regain what we have lost on the ocean.

II.

HISTORY IN BRIEF.

To be monarch of the sea has for centuries been a great national idea with England. Her statesmen have held to that idea through all changes of government, knowing that without control of the carrying trade their little island would amount to nothing; while with it, being the world's carrier and workshop, holding all markets in her grasp, and drawing the gold of all nations to her coffers, England must be most powerful among the peoples of the earth. Hence, to gain that end she has waged wars, freely poured out blood and money, and counted no cost too great; and from the formation of our government down to the present, it was against that determination, which she sought by every means to carry out, that we have had to contend.

ENGLAND'S NAVIGATION LAWS.

In 1650 England passed a Navigation Act, establishing that none except Englishmen should trade and carry to England's colonies; that the ships for this trade must be English built, and three-fourths of the crew Englishmen; that Europeans could trade to England only from their own ports; that the colonies could export only to England; and reserving the coasting trade to English vessels. Under these stringent protective laws England set out. She soon fought two costly wars to uphold them, but in twenty years had quadrupled her ocean tonnage, and in 100 years had crushed the only serious maritime rival then existing—the Dutch. To her rigorous execution of these laws and her consistent adherence to their principles for more than 200 years, England owes chiefly not only her control of the carrying trade, but also her enormous commerce.

A NEW RIVAL.

When the American Colonies declared their independence, England's statesmen were alarmed for her shipping interests. They had jealously watched the colonial growth. They knew that America had abundant and cheap timber, and had already shown skill in building wooden ships. The United States once independent, a dangerous future rival in the carrying trade was recognized, and England's policy shaped to prevent, if prevention were possible, the shipping growth of the new nation.

But the founders of our Republic were not a whit behind England's statesmen in recognizing the necessity of getting a hold on the carrying trade, and did not propose to be dwarfed or driven. As a colonial Confederacy they had already endured England's persecution and oppression of their merchant marine, because they could not help themselves. As a United States government they met policy with policy. Immediately the first Congress was in working order, attention was given to our shipping interests. Discriminating duties on goods and tonnage were imposed, to meet similar duties imposed by England, giving the American ship at least equal footing with the English. It was opposed that such duties would cripple our commerce. Mr. Madison said in answer, that if America should have vessels at all, she should have enough for all the purposes intended: To do her own carrying, to form a school for seamen, laying the foundation of a navy, and to be able to support herself against the interference of foreigners. In this spirit our early statesmen, fearless and defiant, taught the people to depend upon their own abundant resources for the supply of their wants. Alarmed at such prompt, retaliative, and unexpected legislation, which proved how formidable was the new rival, England inflicted all possible annoyance upon our shipping.

OUR NAVIGATION LAWS.

The Americans tried earnestly to bring about a condition of peace and equal rights in trade on the seas. Failing in this, Thomas Jefferson laid down the principle that "if a nation persists in a system of prohibitions, duties, and regulations, it behooves this government to adopt counter prohibitions, duties, and regulations." The American Congress passed Navigation Laws similar to those of England, reserving the coasting trade to our own people, restricting importations to ships American built, manned, and owned, and in every point meeting the laws laid down by England.

The result of the protective laws upon our shipping was wonderful. In ten years so many ships had been built that seven-eighths of our foreign trade was carried in American bottoms, and the China and India trade was well-nigh monopolized by our merchantmen. From 1789 to 1812 our tonnage increased from 280,000 to 1,100,000, or 400 per cent., while England's increased only from 1,500,000 to 2,000,000, or 25 per cent. Besides this, we had built and sold to foreigners some 197,000 tons of ships ; and it should not be forgotten that this was done in twenty-three years, from a start made with a bankrupt treasury, no national credit abroad, only 3,000,000 of people, and a wilderness to conquer. Beginning like that, however, we shot ahead, fought two costly wars—the Revolution for independence on the land, the war of 1812 for independence on the sea—and experienced enough in the way of bankruptcy, poverty and hardship. But through all those trials our earlier statesmen never thought of giving up their policy of self-reliance and going outside either for ships or anything else. It is only now when we have a country developed, 46,000,000 of people, resources superior to those of any other nation, and foreign peoples dependent upon us for bread, that it is proposed we must go abroad to buy our ships, and buy them from our great rival, who has so much at stake to keep all the carrying for herself.

During that period of unparalleled progress, moreover, England went to all lengths short of war to destroy our shipping. She captured 1,660 of our vessels, and confiscated most of them, with their cargoes worth millions of dollars.   She stopped American ships and took from them 6,257 of our seamen, often leaving the ships in mid-ocean without a sufficient crew, under the claim that a person born a British subject was always that and liable to be pressed into England's service.   This claim she enforced, not because she needed the men, but to cripple our merchant marine, and check us in the rapid progress we were making as her rival on the sea.   If she meant to force the young Republic into a war which she knew it could not afford, she succeeded.

THE WAR OF 1812.

It was diamond cut diamond.   Our statesmen were of a temper equal to the occasion and their wrongs.   They met tax with tax, embargo with embargo, wrong with protest ; and when in 1812 they could not answer outrage, they promptly declared war, to settle the rights of citizenship and of liberty on the sea.   They deplored the necessity, for the country was in no condition for war ; but they saw that present suffering was better than abandonment of their rights, and surrender to England of that carrying trade by which our commerce was being extended over the world, our country developed, and our prosperity secured.   The war cost us $150,000,000, besides millions worth of property, and thousands of lives ; but it disposed forever of England's false claims over our citizens, and of open war as a means of driving our ships from the ocean trade.   We won that triumph through the efficiency of our marine and seamen.   Had not the Navigation Laws served us well up to that time?

PEACE'S VICTORIES.

After the peace of 1815 our shipping growth was yet more wonderful, and without parallel in history.   The result was that in 1827, twelve years after a war that left us in miserable financial condition, our own ships were carrying $135,000,000 of our foreign trade, leaving only some $14,000,000, or about 11 per cent., to be carried in all foreign bottoms.   Our sales of ships to foreigners, moreover, had increased to nearly 300,000 tons.   England's shipping interest during this period was as depressed as ours was prosperous.   From 1824 to 1827 the number of her ships actually fell off from 24,776 to 23,195, or 1,581 ships.   Nearly all the British ship owners were losing money.   The feeling aroused in England at this condition of things is well illustrated by the following editorial from the London *Times* of May, 1827:

"It is not our habit to sound the tocsin on light occasions, but we conceive it to be impossible to view the existing state of things in this country without more than apprehension and alarm.   Twelve years of peace, and what is the situation of Great Britain ?   The shipping interest, the cradle of our navy, is half ruined.   Our commercial monopoly exists no longer; and thousands of our manufacturers are starving or seeking redemption in distant lands.   We have closed the Western Indies against America from feelings of commercial rivalry.   Its active seamen have already engrossed an important branch of our carrying trade to the East Indies.   Her starred flag is now conspicuous on every sea, and will soon defy our thunder."

1827 TO 1833.

Things went from bad to worse in England. While ship building was carried on by us with ceaseless activity, it decreased in England from 1,719 ships of 205,000 tons, in 1826, to 1,039, of 103,031 tons, in 1831. Our traders were now in every sea and fast monopolizing the carrying trade.

In 1833 the English merchants and ship owners, recognizing the fact that England had no timber to build her ships at home, and had to bring her timber from America, a distance of 3,000 miles (requiring four ship-loads of material to build one ship, and being compelled to use unseasoned timber and consequently getting ships inferior to ours), petitioned for a repeal of their Navigation Laws.

Parliament ordered an investigation, and for nearly twenty years thereafter the subject was discussed before it was considered wise to make any change of policy. At the end of that time England could well repeal her Navigation Laws, since she had found a new material for ships and had a practical monopoly of what she foresaw, after having made many and expensive experiments, was to be the carrying power of the future; as well as for other reasons which will be shown hereafter. Nor should the fact be lost sight of that it was through those very Navigation Laws that England gained that power and position that enabled her to repeal them with advantage.

The first result of England's investigation was the discovery that Prussia and Denmark could build ships for £8 a ton, France for £11 a ton. The United States, having an abundance of timber, could build the fastest and staunchest ships in the world for from £10 to £12 a ton ; while England, having to import her timber, could not build them for less than £15 to £20 a ton in favored places, and £28 a ton in London. Still, her statesmen saw it would not do to dwarf her industries by letting the cheaper American ships come into her market and under her protection. They knew they could not afford to crush one great interest like that of ship building, with its great employment of labor and vast benefits to the country, in order to favor other interests, but must save and promote them all. Besides, buying ships abroad had been tried sufficiently for them to realize, as they freely admitted, that England must lose her first place on the ocean if compelled to buy ships from us, or to even buy from us the material with which to build them at home. The repeal of the Navigation Laws was therefore refused at that time.

1833 TO 1840.

Our statesmen held equally to their fostering policy, with the happiest results. Our commerce had grown in 1836 to $480,000,000, doubling itself in twenty years. Our tonnage increased in still larger proportion. In 1840 it was over 2,500,000 tons. We had also sold over 400,000 tons of shipping abroad, by this means bringing to the country $25,000,000 in gold, 90 per cent. of which was paid to American labor. Of our carrying trade at this period it has been well said that "its vast profits laid the foundation of the "wealth of the country, and built up its merchant marine with a rapidity "unequaled in the history of the world."

Had not our Navigation Laws up to this time served us well?

England's New Departure.

By 1840 England's statesmen were alive to the danger of her position. They had tried deferential duties and we met them. They had tried war and we beat them. They knew that if they tried those measures again, we would again meet them squarely. Despite all these efforts, they saw that the United States had already grown to be the second carrying nation of the world, had the raw material, mechanical skill, and energy to spare, and promised to become the first. This meant that England was losing her grasp on the carrying trade, and that gone, her commercial supremacy would go also. The time had come when some new policy must be tried. A remedy must be found.

Did those statesmen look to foreign ship-builders for that remedy? They knew well that any remedy that could not be found within their own territory, and which they could not control in peace or war, would in the long run be worse than the disease itself. Did they not judge correctly? It is well worth the while of every American to note carefully the policy England's wise men now pursued, and its results. The many patriotic citizens who are truly interested in having America regain her place on the seas may learn much from the example set by England at this critical moment in her maritime history.

The Iron Steamship.

Success was hit upon in the introduction of the iron steamship. The superiority of iron over wood, of steam over sail, was instantly recognized. After all her costly efforts, after spending millions in war, England found that her hold on the carrying trade was only to be maintained by building her own ships, swifter and safer than any other ships, and placing them on the ocean. As her iron and coal were more developed than those of any other nation, iron and coal were cheaper with her, and within easy transportation to her ship yards. Her statesmen saw that with a powerful fleet of swift iron steamers, England could run away from the wooden sailing ships, secure the world's mails, passengers, and first class freights, and far more than regain what she had lost on the ocean. Such a fleet must be built, and at once.

But how? To construct an iron fleet required rolling mills, engine works, extensive yards. To bring these into existence there must be an outlay of millions of capital. The wooden ship builders could not meet these new and expensive requirements, nor were private capitalists willing to invest the vast sums of money requisite. The statesmen recognized this difficulty, but they were not disposed to be penny wise, pound foolish in a matter of great national concern. The British government, therefore, gave contracts to private ship yards to build iron steam vessels for her navy. By this means the private builders were enabled to establish the great ship yards which have realized for England all that was hoped from them. And in addition to furnishing so liberal direct encouragement at the outset of her iron ship building, England has continued to encourage the builder by having 75 per cent. of her naval ships and engines constructed in the private yards. What would our statesmen say to a proposition that our government follow such an example of encouragement as this? Yet, what was the result? By holding to her policy England made nearly all the other nations dependent upon her private workshops for their ships, naval and marine, thus draining them of their gold, giving employment to thousands of her own working men, and getting back many hundred fold in trade and freight money the sums expended to maintain and extend her control on the ocean.

STEAMSHIP LINES.

This was only one part of England's new policy. Her statesmen knew that to build up a fleet of iron steamships such as they wanted, it was not only necessary to have the resources and facilities, but also to find profitable employment for the ships when built. Their policy worked to their own interest in two ways. By establishing swift mail steamship lines they opened up new markets to their merchants; and the opening of these markets increased the demand for steamships. The English government therefore entered into contracts with private corporations, to pay them such rates for carrying her mails on the ocean as would enable them to run regular lines of fast steamships to the important ports of the world, and induce them to enlarge and perfect these lines rapidly. It was seen that in this way, if in no other, England could speedily leave all her rivals behind.

The fastest and safest ships must of necessity secure the mails and passengers of all countries. Thus England's policy was such that if a merchant in any part of the world wanted to go anywhere by steam, he must first go to England. To that extent she carried her idea of drawing all trade and merchants to herself as the centre; while at the same time giving to her own merchants the advantage of quick communication, quick delivery of goods, and quick collection, which are the life of trade. The more goods the English merchant sold, the more ships were wanted, and the more ships were wanted the cheaper they could be built, since continued manufacturing reduces the cost.

The English statesmen, moreover, did not expect to pay every iron steamer that was put afloat. The fast mail steamships were never intended to carry the bulk of the trade built up by them. They were to bring the merchants of the world to England, because they could bring them more quickly and furnish them better accommodations than vessels running to other countries. And by this means they were to create a trade which should demand a fleet of slow iron steamers, of great strength, cheap to build and run, and requiring no pay. All this these steamships did, more than justifying those who adopted the wise and far-seeing policy of encouragement. These freighters are to the mail steamships what the freight trains on our railroads are to the lightning express. But those who oppose encouragement by our government urge that if you pay one class of ship for mail service you must pay another, and they show the enormous expense this would involve. This is not true. The slow freighters always take care of themselves, and the sum paid to the fast ships is nothing when put beside the revenue obtained as a result of such encouragement.

VALUABLE RESULTS.

The experiments in securing swift mail service were attended with most gratifying results, and by her new policy, which cost her a mere nothing in comparison, England gained greater control on the sea than she ever had during the 200 years in which she had waged wars and expended millions upon millions to establish her mastery there. In 1840 Samuel Cunard was paid $450,000 a year to carry the mails semi-monthly to Halifax and Boston. It was not long before the trips were made weekly, and the pay increased to $725,000. In five years such were the returns to England from this service in trade and freight moneys, enriching her people and filling her treasury with gold, that mail lines were established to China, Japan, the West Indies and Mexico, and England was paying these lines some $4,000,000 a year, already realizing, what her shrewd statesmen foresaw, that they were not only mail carriers but Trade Pioneers, and repaying a thousand fold what they received.

ENGLAND'S SUCCESSFUL STRATEGY.

How did our statesmen meet England's new policy of paying her way into the markets of the world? Had a cannon ball been sent into one of our merchantmen, had a duty been imposed upon our ships, does any one doubt what our answer would have been? We were not so ready to meet diplomacy. At first, it is true, some attempt was made to reply in kind, and the old disposition to contest every foot of the way was manifested. But the attempts were feeble and inadequate and presently given over. And all subsequent efforts to hold our place on the ocean by putting mail contracts against those of England, by paying for mail service on the sea as we paid for it on the land, or by giving equal advantages to our ships in the foreign trade, were crushed out by the misuse and twisting of the word Subsidy, which was fastened to ocean mail contracts as an "execrable shape" to frighten a great people from advancing their own interests, and was rendered by shrewd manipulation as odious as bear-baiting or court frippery to the Puritans, as frightsome and terrible a thing as witchcraft to the early New Englanders.

1845 TO 1860.

The consequence was that from the introduction of her new policy England began to gain on us. While this was not at once apparent enough to cause the alarm necessary to awaken our people, the new policy laid the foundation which enabled England to seize the opportunity that came not many years afterward, through our war. But for that opportunity we, with the great capital involved, would never have allowed England to run us out on the ocean.

The growth of our wooden sailing vessels had continued. In 1850 we had a tonnage of 3,335,454, an increase of over 200 per cent. since 1815. Our coasting trade, from which foreign ships were excluded, employed 1,900,000 tons of shipping, an increase of nearly 400 per cent. in the same period. All this while England was paying $4,000,000 a year to her fast steamships, which were intended to drive our clippers from the sea. Yet, under our wise laws our tonnage increased to 5,350,000 in 1860, and of our total foreign trade $437,190,000 was carried in American bottoms, against $160,057,000 in foreign ships. It must be borne in mind, however, that our growth was almost wholly in the wooden sailing ship, while England's was in the improved iron steamship. But up to this time, in face of a persistent policy to break us down, had not our Navigation Laws served us well?

WOOD AGAINST IRON.

In 1860, through her policy of encouraging the building of an iron steam fleet, and of paying remunerative wages for ocean service, England had 156 ocean steamers, of 210,000 tons, engaged in the carrying trade. The United States, through our failure to meet England in this wise and peaceful policy as we had met her in every other, had only 52 ocean steamers afloat, of 71,000 tons, in both the coasting and foreign trade. The rest of Europe outside of England had 130 steamers, of 150,000 tons, principally built by England. This left us with 52 steamers, many nearly worn out, against a European fleet of nearly 300 steamers, or six to our one. In the carrying trade from this country to Europe the foreigners had 31 steamers engaged, and we had five. England was paying $5,000,000 annually for the carrying of her ocean mails; France was paying nearly $4,000,000 for ocean service, (large portions of it to lines running to New York), she recognizing the necessity of meeting England on her own ground. Our government was doing nothing to enable our ship owners to compete with their rivals. Their ships having greater speed than ours, they commenced taking the best part of the business, and every attempt to prevent the total loss of our carrying trade was overwhelmed by that fateful howl of "Subsidy." We had been beaten by England in diplomacy, and surrendered to strategy what never could have been wrested from us by force.

III.

THE REBELLION.

Now came our civil war and England's opportunity. After her 200 years of steady struggle, by any means and at any cost, to destroy her rivals in the carrying trade ; after her policy of war, oppression, and strategy pursued from the beginning of our national existence to crush us out on the ocean, as has been described, in what mood do you suppose England greeted the deadly struggle within the borders of the only formidable rival left? Would she try to bring the warring sections to peace, or be disposed so far as possible to increase their hatred and divisions? While we had unquestionably many warm friends in that country, who deprecated both our unhappy condition and the attitude assumed by their own government in its unconcealed sympathy for the Southern cause, it must not be forgotten that the party which favored the South was largely predominant, and since England had never hitherto stopped at the morality of her acts on land or sea, was it likely she would now hesitate to strain her conscience a point in her own interest? For the North and South to ruin each other meant double business for her : on the ocean she could grasp our carrying trade ; on the land, in the South she saw a chance to secure for herself a cotton and bread producing country, as well as a market in exchange for her manufactured articles.

England would have been only too glad to have kept the South as just such a country to trade with. But the result of the war, in maintaining the Union, which England would only too gladly have seen broken, put an end to that idea. The people of the South have awakened to the fact that their section has coal and iron, as well as cotton and bread, and that in her place among the great family of States she can develop all her resources and become self-reliant as she never could have done out of the Union, with the pressure England would have brought to bear on her. The South to-day will hardly relish that foreign interest in her affairs that would have kept all her resources undeveloped except her cotton and bread. Experience has taught her that her weakest point has been and still is the want of mechanical power to develop her great resources, in which no section of our country is more favored.

AN INTERPOLATION.

The recent realization by the Southern people of the advantages of their position for certain manufactures is worthy of note. Gen. Hooker, one of Mississippi's representatives in Congress, in some remarks made at the recent meeting in New York to promote the movement for a World's Fair in 1883, said the South and Southwest felt great interest in the movement because "such fairs make us feel, by bringing the people of all sections "together, that we are one country, having an identity which overlooks "sections, and makes us indeed an united and indivisible people." He said further that the South had something to show in this Fair. She had to show that her people realized their advantages at last, and had already made such advance in the manufacture of the coarser textile fabrics (for which they had the best facilities and most profitable place in the world), as the North little dreamed of. A year ago, at a meeting of manufacturers in Boston, a gentleman from the Southwest who had just opened a cotton cloth factory close to the plantations, served warning on the New England cloth-makers that they must look out for sharp competition; and they replied, "We know it; we know you have the cheapest and best location, "and it has always been a marvel to us that you did not find it out years "ago." So Gen. Hooker gave notice that the South was alive to its interests, manufacturing as well as agricultural.

As showing, also, the keen interest felt in Congress on the subject of our carrying trade, at this same meeting Senator Windom, of Minnesota, took occasion to say, after declaring that our prosperity would continue just so long as we could sell what we produce, that our great danger to-day is over-production, and our great want new markets for our surplus products:

"The time has come, sir, when we should no longer depend upon our chief competitor in the world's markets to carry our products to those places where they are needed. In the sharp competition of the future we must regain our lost carrying trade, and boldly challenge our great competitor for the mastery of the sea. We have suffered long enough under an unwise policy. When competing countries are using every means to draw the merchants and trade of other nations to themselves, we must not sit quiet."

These remarks were received with an outburst of applause. Gen. Hooker also alluded to this subject, saying that although he represented an agricultural section, yet every section and class of people were interested in our carrying trade ; that it was a shame that under an unwise policy we had been reduced to the fifth carrying nation, when we ought to stand the first on the earth. He hoped there was no man in America who would not give his support to place us where we belong on the ocean. He knew that, while his section was agricultural, it was as much concerned in the revival of our carrying trade as were the manufacturing centres, and that the shipping interest was national in its influence and effects.

A DESTRUCTIVE NEUTRALITY.

· To return to 1860, under the guise and cover of neutrality what did England do in our crisis ? She fitted out armed cruisers and blockade runners for the Confederacy. There went out from her ports the *Alabama* and *Shenandoah*, which came among our wooden ships like wolves among a flock of sheep. Do you imagine that she sent out these destroyers because of any love she bore to the South ? They were simply instruments used for the furthering of her policy to annihilate our carrying trade. Perhaps some will ask contemptuously what the *Alabama* and *Shenandoah* or a dozen more ships like them could have to do with the carrying trade of a great nation? Let us see.

OUR FALSTAFFIAN FLEET.

Our ships were principally wooden sailing ships. Our steamers, what few we had, were slow wooden side-wheelers, and their large wheels being high out of the water were easily disabled. We had nothing to match the iron cruisers. We had no navy sufficient to protect our coast, let alone our merchantmen on the high seas. Why, it took our whole steam fleet, naval and merchant, to look after the English built blockade runners. For defense and transportation the government absorbed more than a million of our tonnage, and might as well have absorbed the rest, since it was left defenceless, and the risks were so great that merchants could not afford to trust their cargoes in American bottoms. Our financial condition, moreover, was such that a large portion of our ships and cargoes had to be insured in Great Britain, at such high rates as to be entirely destructive to profits. The cruisers did inestimably more damage than merely destroying the 100,000 tons of our shipping with which they were credited. They rendered the ocean so unsafe that our merchants had to do what England foresaw and desired—put their goods into English ships and tie their vessels, what they had left of them, to the docks.

So, by this policy the carrying trade of the North Atlantic was left for England, because she was the only nation that could build the ships to take it, and its vast profits poured into her rapacious pockets. Thus she drained us of our gold by the millions when we were helpless. Is it not surpassing strange that we should let her continue to drain us when we are no longer helpless, but are in position, and only need a wise policy, to give the benefit of these hundreds of millions to our own people?

RICH REWARDS.

The result was that England doubled her ship building, from 208,000 tons in 1861 to 462,000 in 1864. And as the ships were principally steam, and each ton of steam is equal in capacity to three tons of sail, the actual increase was equal to 752,000 tons. England was too wise to attempt a recognition of the Confederacy. Aside from fear of European complications that would result, her statesmen knew the opposition they would have met in such a movement from the masses of their own people, led by such staunch friends of the North as John Bright. But the Southern sympathizers, who controlled the government, did everything short of that recognition to help break up our carrying trade. And the policy pursued under the artful guise of neutrality did the North more damage than open enmity, and would, it was thought, cripple the nation for a century. But our recuperative power was too great, and the spirit of our people too independent, for such a result. Nevertheless, through the opportunity opened by our war, England accomplished the purpose intended when she entered upon her policy of building an iron fleet and paying it to render her ocean service. When the war was over, her flag covered the ocean. Our marine lay crushed and helpless, and England had almost as complete possession of the North Atlantic with her iron steamers as we have of our lakes and rivers. But for the war, I do not for a moment believe we should have allowed England thus to get away our carrying trade. With the large capital involved, we should have devised some means to meet her policy. But neither could she have taken it by aid of the war had not her policy previously furnished her with the means to seize just such opportunity.

PALTRY INDEMNITY.

For all this damage and destruction and the millions she gained from us, England paid us a paltry $15,000,000 for damages. This was like a man helping to destroy your place of business at a time when you could not rebuild it or punish him for the crime ; fitting himself out with a stock of goods, taking your place in the market, realizing $100,000,000 out of the business, and afterward compromising by paying you a small portion of the profits, but still holding the position thus treacherously obtained. And when you wish to commence competition with him, you are advised to depend upon your destroyer for the rebuilding of your store, and thus pay him further profits. That is an exact illustration of the situation we are in respecting our carrying trade, and of the advice given by those who cry out for the repeal of the Navigation Laws. But for those Laws indeed, there would have been no idea of trying to re-enter into competition on the ocean.

WHAT MIGHT HAVE BEEN.

Now, suppose we had met England in 1840 as we met her policy at all other times prior, and had in our turn encouraged the building of iron steamers to equal extent, by equally liberal pay for fast mail service, and by other encouraging legislation. Suppose we had, as a result, not only developed our iron and coal, and stimulated all branches of industry, but also built up a similar fleet of 150 iron steamships, ready to be summoned to government service and defense when the war broke out. Why, with 75 of these fast steamers we could have both protected our coast and blockaded every Southern port, leaving the other 75 ships to carry on our business and take care of outside invaders. In that case we might have forced England to pursue a different kind of neutrality, and the *Alabama* and *Shenandoah* would have had no existence. Our war might have been ended within a year, and thousands of lives and millions of dollars saved to our country. What a tremendous sacrifice did we make then for want of a wise policy twenty years before in regard to our shipping interests ! Had we appropriated twice as many millions a year as England did to encourage building such a fleet, would not this expenditure have been saved to us many times over when our emergency came ?

IV. ` ,

AFTER THE WAR.

What was the condition of affairs when peace was restored? The war had stimulated internal development. The millions of capital withdrawn from shipping by American merchants had been put into railroads, telegraph lines, and factories. There was no chance to profitably invest capital in competition with England on the ocean, where there was no protection for it, and naturally this capital sought investment where it was protected.

The agents of American shipping houses abroad had been called home, and to-day we have fewer American representatives of our shipping houses in European ports than we had fifty years ago, while their places have been filled by the representatives of foreign ship owners. The loss of these representatives is a severe one to us because they were constantly looking up trade and a market for exchange. They always had something to sell in other markets that would make a return cargo for their ships, and were valuable pioneers of trade. No attention had been paid to our shipping interest, and England was in almost undisputed possession of the vast business which our internal troubles had thrown within her reach.

If, immediately after the war, our statesmen had recognized the imperiled position of this great interest, and adopted a policy of encouragement calculated to restore us gradually to our former position as a carrying nation, who can doubt that millions of dollars annually would have been saved to the country ; millions more paid to American labor, instead of going as they did to support foreign peoples; the day of specie resumption hastened, and very much of the distress consequent upon the hard times averted? The trouble was two-fold: 1. That in the development of two of the great sources of national prosperity—Agriculture and Manufacture—we nearly lost sight of the third equally great source—Commerce; and, 2. That the efforts made by our merchants and ship builders to restore our carrying trade, (which is the life and promoter of commerce), were both discouraged and thrown under odium by the cry of Subsidy, instead of being appreciated and encouraged as they should have been.

OUR SORRY CONDITION.

When the attention of the people began to be drawn to the carrying trade, what we had left of it, it was found that of our remaining tonnage a small proportion was composed of wooden side-wheel steamers, almost worn out and of little use ; the balance of wooden sailing ships, many of them also old and comparatively worthless. Although during our war the revolution on the ocean—steam for sails, iron for wood, the screw for side-wheels, the compound engine for the ordinary—had been completed, yet at its close there was not an iron screw steamer, nor one with a compound engine, under our flag. Nothing left us but wooden sailing ships to compete with England's fleet of modern iron screw steamers! We had no yards established for the building of iron ships. Our rolling mills were not in condition to make the shapes of iron necessary, nor had we much skilled labor in that direction. Our currency was at a discount of forty per cent., and we had a tax on the ownership of vessels ten times greater than Great Britain's. Added to all this, England had the powerful advantage of possession. As business men you well know the difficulty of organizing capital to buy ships or anything else in any market for the purpose of competing with capital already organized and invested, and especially when the surrounding circumstances are all against the new organization.

THE CURSE OF AGITATION.

Worse than all else, our ship owners and builders were not even left free to see what they could make out of the unpromising situation. Look at what was now done, following out the exact line of England's policy as we have traced it from 1789. England knew that this was the only country that had the natural resources necessary to compete with her in building the iron ship, which was to be the ship of the future; hence the only country that could interfere with her practical monopoly of the carrying trade. She also knew, from bitter experience, that given the chance we would again become her lively rival. But to do this we required a vast outlay of capital in iron ship yards and rolling mills, a capital which could not have been raised in England when she changed from wooden to iron ship building but for government aid. She knew that nothing could so surely prevent the American merchant and builder from investing this capital as the presentation of a free ship bill in Congress, and the continued agitation and pressure of that bill. Thus from 1865 to 1870 a free ship bill was kept hanging over the heads of our capitalists who were disposed to invest in the carrying trade. By this means all efforts to rebuild our shipping interest were made useless until Congress in 1871 defeated the schemes laid to pass the free ship bill. That bill was so plausibly presented as to secure the support of many honest and conscientious men, who had not the time or means for a thorough investigation of the subject, and who took their tone from a portion of the press which was hammering away on the ideal free trade doctrine. The chief and most taking argument put forward by those most actively concerned was, that if we could buy ships in England as cheap as the English owners, we could run them in competition with the English lines. The fact that the slight difference in the original cost of a ship was the least part of the difficulty in the way of our ownership, and that the real hindrances were those of taxation, high capital, &c., as will be shown, was carefully kept in the background. By the defeat of the bill, the encouragement of protection for their interests was given to American merchants and ship builders; and from that action we date the start of iron ship building in this country. Further on will be shown the progress we have since made.

V.

FREE SHIPS.

This brings us to the important question of free ships,—a question that is never long allowed to drop out of the notice of Congress. Let us look at it a moment, and see if we can get at the truth of it, and find who are most interested in its agitation.

Permit me to say, in the first place, that I do not question that there are thousands of high-toned, conscientious men in this country, including many representatives in Congress, who believe that in the repeal of our Navigation Laws rests the remedy for our depressed carrying trade. I would not for an instant doubt the honesty of this class of our citizens, who advocate what is speciously called the "buying of cheap ships," under, as I believe, mistaken ideas as to the real meaning and effect of such a measure as they propose to help pass. In whatever I may say here of the free ship advocates, I wish to entirely exclude this class of honest and conscientious men. It is to them I appeal for candid consideration of all sides of this subject, and to them that I present my views, as the views of one who has grown up in the shipping business, and given his life to the study of its interests; of one who holds that America from her position and resources ought to be, and must be, the ruler of the sea. We have a right to that place because we have more goods to float on the ocean than has any other nation. But to be the ruler, we must in peace and in war alike be able to supply our own wants in ships. History teaches that no nation that did not build its own ships ever controlled the ocean. The reasons are patent why none but the ship-building as well as the ship-owning nation ever can hold that control.

VARIOUS ADVOCATES.

There are, however, other classes of men who join in the cry for free ships—some of them honest but ignorant, others intelligent but dishonest. There is one class composed of men who do not care to become informed, and never go into the matter beyond saying: "My policy is to buy in the cheapest market and sell in the dearest." Nothing can draw from them anything but that. You try in vain to show them the truth that the cheapest market in price may be the dearest in fact for their buying, and the dearest market be anything but the most profitable for their selling. They are chained to the post of what they consider an economical axiom, and cry out, "Let us buy ships where we can buy them cheapest," though as a matter of fact they have no idea of buying them anywhere.

Another class includes those who think it the genuine American idea to slap their hands on their pockets and say, "I believe that I should be allowed to buy whatever and wherever I like, so long as I pay for it." These men, also, have no idea of buying ships, but cry out for the repeal of the Navigation Laws merely as the result of their devotion to what they regard as the principle of our liberty.

Then there is a class of men who are disposed to take a liberal view of the question of mail contracts, and believe in wise government encouragement. But they say that, on the ground of rigid economy, they want the chance given to buy cheap ships abroad for this service. Now, suppose the government proposes to give the owners of these mail ships in the form of mail contracts three per cent. on their capital invested. Let the cost of the English ship be twelve per cent. lower than that of the American. Aside from the important considerations of labor, tax-paying, &c., all paid at home, and the advantage in point of national defense, will these economists figure out how much the government would save in its three per cent. on that twelve per cent. difference ; then figure how much we should lose on every ground of national interest, and see which is the true economy in the end, to build our ships at home or buy them abroad.

Still another class are men who are interested as the nominal owners and managers of the old wooden sailing ships. The ownership principally resting with the builders in Maine and elsewhere, these men had little money invested. These ships are rapidly disappearing, and the men who ran them and thus gained acquaintance with our carrying business, would like well to run iron ships for foreign owners on the same terms of small investment, and in this way use their knowledge to good advantage, while foreigners furnished the ships.

These classes may be perfectly honest in what they do. But there is still another class, of wholly different character, composed of agents and others interested in carrying out England's policy to prevent our shipping growth, and so cut off at the root the possibility of our again becoming her rival in the carrying trade. These men, whose selfish and dishonest purposes will be shown plainly, shrewdly use the other classes of our people to further their own designs, and under the specious guise of free trade they draw many whose intelligence should lead them to support the interests of their own country rather than those of any other, but who seem blind to the truth of this great subject.

It is to this latter class of designing advocates, who will be found to have no money invested in ships, and no money to invest, but to be working in a foreign interest, that I refer in the arguments which follow.

A FALSE REMEDY.

To revive our carrying trade, exclaim these men, repeal our Navigation Laws; then all will be right. Let us see about that. We have shown how our shipping, under the protection of these laws, grew from 1789 to 1860 without parallel in history. We had no check in our triumphant progress on the seas until we failed to meet England in her policy of encouraging iron steam ship building, paying for ocean mail service, and other wise legislation. Even that policy would have been insufficient to entirely break our hold on the North Atlantic carrying trade had not our war distracted public attention from the subject, and our government subsequently failed to give protection to the capital of our merchants invested on the ocean, thus forcing capital into other channels. Up to that time was there any fault to be found with our Navigation Laws? To them we owed the foundation of our commercial prosperity, our success in establishing the freedom of the seas by the war of 1812, and our rapid strides to a foremost position among the carrying nations of the world. In every requirement had they not served us well? What end to the advantage of our country is to be gained by their repeal now?

ENGLAND'S REPEAL.

"Oh," says one, "because England repealed her Navigation Laws. If it was good for her it will be good for us." But when and why did England repeal those Laws? When, for reasons already shown, England could no longer compete with us in building wooden ships; when she had found her remedy in iron, and after ten years of costly experiments and by a policy of encouragement had proved that she could make iron ship building and running a success; when through her swift and superior steamships she was sure of getting the mails, passengers and fast freights of all nations, she then repealed the Navigation Laws which for 200 years had stood her in so good stead as to make and keep her the first carrying nation of the world, and to build up for her that maritime greatness that enabled her to do without them. Because then there was no danger that her merchants would wish to buy our wooden sailing vessels; and by that repeal she threw into active competition with our sailing fleet the wooden ships of the Dutch and other peoples, who could sail ships cheaper than we could. This cheap competition helped to break down our carrying trade; while England knew that with her iron steamers she could eventually vanquish the wooden fleets of all countries. Into what a mighty power has England converted her coal and iron, by her statesmanlike policy; and though she has expended immense amounts to achieve this success, what wonderful results she has to show, and how many times over she has been repaid! Is it not worth to her far more than has been spent, that to-day she has half the sea-going tonnage of the world under her flag, carries the products of all nations, and builds ships for nearly all? If she can but succeed in closing our ship yards, she has nothing further to prevent her reaching the position which she has for years struggled to occupy, when she can say defiantly, "My coal, my iron, my workshops do and shall control the ocean." Are there any similar circumstances which would make it wise for us to remove the last protection and the only one given our ship owners and builders against such a powerful carrying force as the English builders and owners have been in every way liberally aided to establish?

WHEN WE CAN PROFITABLY REPEAL.

When we can build the iron ship as cheap as England can (and we shall be able to do it when further developed, if our industries are not crushed by legislation) ; when we make the taxes upon our ship owners as light as those of foreign ship owners ; when we study the interest of American ship owners as England has studied that of hers ; when American merchants are aided as hers have been to get control of the carrying business, and can defy competition, then we may profitably discuss the repeal of our Navigation Laws.

What would England's position on the ocean be to-day if the enormous amount of capital she has invested in her shipping had been paid to foreign countries for ships and repairs? Was it the right to buy ships wherever she chose that has given her the place she now holds? Or was it her ability to build her ships at home, out of her own resources? France and Germany furnish the answer, as will appear further on.

WHY WE SHOULD NOT NOW REPEAL.

We should not repeal our Navigation Laws because they are not at the bottom of the difficulty at all, while they are of vital importance in any true effort made to remedy that difficulty. The truth is, that it is not the slight difference in the original cost of the American ship that keeps our merchants from buying. The real reason is, that they cannot afford, under our laws of taxation, rates of capital, prices of labor, and the foreign competition, to own and run ships. They were driven out of the carrying trade under peculiar circumstances, and while those circumstances lasted the business was taken possession of by a people who had the assistance of a government which made that trade a national question. The start must be made against such odds as these. The dishonesty of the interested men who actively lobby for the free ship bill is shown by the fact that they never see anything in the way but the Navigation Laws.

It may shed some light upon the matter to consider, first, under what circumstances our merchants who truly desired to revive the carrying trade would have had to buy ships, if they had bought them at all, at the time this free ship bill was first pressed upon Congress. The same circumstances, with the exceptions of a special war tax and a depreciated currency, still remain an effectual barrier to the ownership and running of vessels under our flag. If it shall appear, then, that it was impossible for our merchants to buy or own profitably, and therefore nonsense to say they wanted to buy abroad, some other interest than that of our ship owners and merchants must be looked for behind this movement.

GREENBACKS AND GOLD.

Between 1865 and 1870, while the free ship bill was vigorously pressed before Congress, the average premium on gold was forty per cent. The American merchant, in consequence, must have paid forty per cent. more for an English ship (to buy which, of course, he would have to convert his greenbacks into gold), than the English, French, or German merchant, whose currency was at par. No business man would think of throwing away his money in such competition as that. Moreover, if the American had confidence in our government he believed that our currency would again be brought up to par with gold, and that he would save this premium. He certainly did not want to buy ships abroad at such disadvantage, hence he was not the man who desired the repeal of our Navigation Laws.

OPPRESSIVE TAXATION.

During the war and for some time after there was, independent of all other taxes, a war tax of two and a half per cent. on the construction of ships and machinery. This was not much like England's encouragement to her builders of iron ships.

With the interest on American capital from two to three times higher than that on foreign capital, how was competition possible? The foreign companies already organized, moreover, were both in possession of the business and had in operation all the ships then needed—two advantages in themselves which you will readily appreciate.

The carrying trade on the ocean is conducted similarly to that on the land—by great corporations in the form of stock companies. The capital invested in ships by an American corporation in New York is taxed at the same rate as houses and lots, two and one half to three per cent. without regard to the profit or loss on the investment. The English corporation owning ships in London and engaged in the same trade is taxed upon its capital invested *only one per cent. on the net profits.*

If the American line earns nothing, if its ships are tied to the docks, the tax on the total value of its property must be paid just the same? If the English line earns nothing, it pays nothing in tax. What chance for competition does that kind of taxation leave the American corporation? Is it not a direct discrimination against us in the foreign trade? Has the advocate of free ships ever pointed out this great difficulty in the way of American merchants?

SURPRISING FIGURES.

To illustrate: Take two lines of steamships, one European, one American, say both semi-monthly lines between New York and Liverpool. These lines own ten steamships apiece, costing each line an outlay of $7,000,000 capital. The tax on the property of the American company in New York, at two and one-half per cent., would be $175,000 per annum. In Liverpool the similar ships, costing the same amount, would be taxed one per cent. on the net profits of the English company.

Now suppose the net earnings of the English company to be eight per cent. on the $7,000,000 capital, or $560,000 a year, one per cent. on this would be $5,600, or the tax on the English company ; while under our laws of taxation the American company, whether its net earnings were eight per cent., or four, or nothing, must pay $175,000. Assuming the net earnings to be the same, reduced to tabular form the showing is :

Lines.	Capital Invested.	Net Earnings.	Taxation.
American	$7,000,000	$560,000	$175,000
English	7,000,000	560,000	5,600
Discrimination in tax against American Line			$169,400

Compare the difference in annual taxation with the difference in the cost of original construction, and which difference is more likely to be the one that prevents our merchants from owning ships? It is not, then, that they cannot afford to buy ships, but that they cannot afford to own and run them against such discrimination. It would seem to be hard enough that the interest on the American capital should be seven per cent., while that on the English was only three, without adding such a tax. This tax on the property of American steamship companies in no way benefits the country, for it prevents the investment of capital in carrying lines to compete with those of England and France. If laws were passed reducing this unjust tax, and removing other discriminations, it would be a step in the right direction in aiding the starting of American lines ; and with light taxation more money would be collected in revenues than at present is collected under the high taxation, to say nothing of the immense benefits to both our industries and trade.

But these figures seem incredible. Can it be possible that our government not only refuses to meet England's policy of encouraging her carrying trade wherever it requires encouragement, but actually places so impassable barriers as these in the way of our regaining a place on the ocean? Can it be that we have no policy in relation to our carrying trade save one of destroying it by taxation and other unjust laws? And why and how is it that the men who pretend to so deep a concern about that trade never mention these difficulties, but simply and always keep up their old cry, "Repeal the Navigation Laws." What, pray, have the Navigation Laws to do with these serious facts just shown, which are the real obstacles?

FURTHER DISCRIMINATIONS.

But these are not all the advantages which the English ship owner possesses. The English company in the foreign trade is allowed to draw all its supplies out of bond duty free—an important item in favor of that company, since the large lines in the North Atlantic trade carry thousands of passengers, and use large quantities of spices, teas, wines, segars, &c. The American company has no such privilege, but must buy from the grocer who has to pay the duty on these supplies. Apply this to a land interest. How long could the proprietor of one of our great hotels, having no privilege from government, run his house in competition with another who was given the advantage of obtaining his imported supplies from the Custom House free of duty?

Again, an American steamship cannot clear from any port, foreign or otherwise, where an American consul is stationed, unless she carries the American mails for the mere postage. The English ships will not carry England's mails unless fairly compensated by the government, and is free to do as her owner likes.

Surely our policy, so far as we have one in regard to our carrying trade, could hardly suit England's purposes better if her statesmen themselves had framed it for us. Would it not be at least wise to examine the liberal concessions made by England, France, and other nations to their ship owners, and to investigate carefully this whole subject before we repeal those Laws which have for nearly a century served us so well, and which are in no wise responsible for our present low condition as a carrying power?

THE ATTEMPT OF '71.

All public men know the desperate effort made before Congress in 1871 to repeal our Navigation Laws. The reason for the special pressure at that time will give an idea of the parties to be directly benefited by such repeal. The French and German war had just begun, and it was expected that all Europe would be engaged in it. Hence, foreign ship-owners were eager to get the protection of our flag by "whitewash" sales, bogus mortgages, and false titles. They could then go on with their trade as usual, and as soon as peace was declared in Europe, there was nothing in the law to prevent them from putting their ships back under the old flag, where they could be run cheaper. We might, it is true, gain some increase of taxable property while they remained with us, but there never would be an honest American owner while ships could be managed with less taxation under another flag than under our own.

VALUE OF OUR REGISTRATION LAWS.

Much specious complaint has been made because a large part of the vessels which hauled down the American flag and put themselves under the protection of England during our war were not allowed to return to an American registry when peace came. There were two strong reasons why they were not allowed that privilege. One reason was that it involved a great principle and would set a bad example. A citizen could in time of war with equal right claim it his privilege to transfer his whole movable property where he would not be called upon to bear any burden of the war, thus taking millions of taxable property out of reach of war taxes, and leaving the burden heavier for those who stood by the country in her hour of peril. The second reason was that this prohibition was one of the most important advantages conferred by the Navigation Laws upon this country, with its immense coast, enabling it to get along with a small navy and thus save millions of dollars annually, by holding our whole merchant marine at command. "If you pull down the flag in the hour of the nation's necessity, you cannot hoist it again." That is what the Navigation Laws plainly say, and thus they nail the flag to the mast. The statesmen who framed those laws recognized fully the value of that principle, which in peace gave us protection, and in war securely bound to us our marine. The policy is sound to-day, and will be maintained by every citizen who believes that all should share the country's burdens in war as they share its blessings in peace.

It is worthy of note in this connection that in the war not one of our steam vessels was transferred. The ships transferred were a large number of old sailing ships, most of them well worn and useless in the war. To bring them back now would be as serviceable to our shipping interest as to bring back the old Dutch galleys. We have too many of that class of vessels now. This argument, however, shows how little some of the interested persons who talk on this subject appreciate what will really help us to regain our carrying trade. We do not want to buy old wooden ships, or old iron ones, or any kind whatsoever. If we are to regain what we have lost we must build our own iron steamships, the fastest and finest in the world, put them on the ocean and keep them there by removing the obstacles now in the way, and giving them a fair chance to compete with their already powerful and established rivals.

VI.

DIFFERENT FORMS OF THE BILL.

This free ship bill is, in one dress or another, continually brought before Congress. Whatever the form, you will find that the real object is to keep us dependent upon England for our ships, and that the movement begins from without, not from those of our citizens who are truly interested in buying and owning ships.

I. The bill is presented in a form affecting the coasting as well as the foreign trade. But in the coasting trade, what American company wants to invest in ships and run them under an American registry, even if they could buy them as cheaply at home as in England ? They would save money by owning ships under the English flag, for they would then have the benefit of low taxation and the other advantages mentioned. And I am inclined to think that many merchants in New York and Brooklyn would like to place their stores and property under the British flag, provided they could be taxed in the same proportion as exists between British and American ships. No doubt the English, who find our ocean carrying so profitable, would like to have the Navigation Laws repealed so as to allow them to do our coastwise and inland carrying also. But do we wish to legislate for foreign interests ? Shall we keep our coasting trade, or allow that to go along with the foreign trade we have lost ?

II. Defeated in this form, the bill is put in another, touching the foreign trade only. As to this, it is clear that no American company organized with a capital of millions would buy, either abroad or at home, ships to run from New York to Havana or Mexico, while it was debarred from running those ships at the same time to San Francisco, New Orleans, Charleston, Galveston, or Savannah. If a bill were passed in that form it would certainly throw the whole coasting trade open to competition, for it is very doubtful whether the Courts would sustain any law preventing an American citizen from carrying his property from one State to another. No American company would invest capital in ships and surrender that right, consequently it is evident that the demand for the passage of such a bill does not come from American citizens who honestly want to own ships.

III. This failing, the bill is presented as a free ship bill, with free materials. This is the most glaringly deceptive of all the forms. The ships under discussion, which the free ship advocates wish to sell to Americans, are iron steamships. Wooden ships are built here now cheaper and better than ever before, and cheaper and better than anywhere else. But we have nearly enough of that class of ships already. The ship we have to compete with is the iron ship built on the Clyde, convenient to where the iron itself is produced. Now, the advocates of this bill try to convince the ship builders and owners of America that it is to their interest to buy the iron on the Clyde ; pay freight, commissions, and insurance thence to Boston, New York, or Philadelphia ; ship it from these ports by rail or otherwise to the different interior points where it may be needed ; and then compete in the construction of the ship made out of this raw material with the Clyde builder who has the iron at his hand. Nevertheless, the decrease of our ship building has been frequently attributed to the high duties imposed upon the raw material, and great stress has been laid upon the point by those who have labored in and out of season for the repeal of our Navigation and other protective laws. The arguments employed to make out a case are familiar to you. But they have only been a source of amusement to practical men who understand the subject of ship building. To prove what there was in the cry for free materials, moreover, a law was passed in 1872 giving to our builders the right to import much of the material that is used in the ship duty free, when the ship was to be employed in the foreign trade. This was considered so impractical by the ship builders, however, that I cannot learn of a single man who ever availed himself of the provisions of that law.

In view of the result of that experiment it would seem as though the raw material argument ought to be dropped. But let us look at it a little further, and see whether the advocate of this form of the free ship bill is informed as to the subject he is dealing with. Is he not in any case, advocating the cause of somebody else than the American owner or builder of ships? Where is the advantage of getting this raw material here free of duty? A 2,000 ton ship requires 1,000 tons of raw material, as the shapes could not well be ordered from a foreign mill. The 1,000 tons of raw iron come here at say 1¼ cents per pound. This must be worked up by American labor into a complete ship, including engines and boilers. This labor—the cost of which is to-day the only reason why we cannot build iron ships as cheaply as England—will bring the price of the iron up to at least ten cents a pound. Where is the protection, then, for the 8¾ cents, or what may be called the cost of American labor.

Between 1820 and 1830 England's builders tried the experiment of buying the timber for wooden ships from this country, but they found the cost of transportation, insurance, and other expenses entailed by this process ruinous to their interests, as the report elsewhere in this paper of the condition of England's shipping interest in 1827 clearly shows. It is not the raw material but the finished product from it that demands attention and fostering, and the advantage of the protection given to that accrues to the American workingman who produces it.

VII.

ENGLAND'S INTEREST.

There are many reasons why the English ship builder and merchant desire the repeal of our Navigation Laws. That repeal would crush out the start which we have made in iron ship building despite all disadvantages and discouragements met where we should have been encouraged and favored. How quick England's people were to appreciate this start, and the watchfulness with which they note our doings (no doubt keeping closer track of our progress than do our own people), is shown by the following quotation from the London *Telegraph* of May 2, 1877 :

" Twenty-five thousand workmen employed in the ship building yards on the Clyde were locked out on Saturday conformably to the decision of the masters, who have unanimously determined to resist the demands recently made by the men for higher wages. We trust this unhappy labor dispute may not result in driving away another important industry from our shores. If recent intelligence from the United States may be believed, British supremacy in the construction of iron vessels appears now to be seriously threatened by American competition. It is affirmed that the iron ship builders of Pennsylvania and Delaware are at this moment building better ships than can be produced on the Clyde; and if that assertion can be substantiated, an industry in which we have been accustomed to regard ourselves as unrivaled is unquestionably placed in great jeopardy."

'That repeal would ruin our enterprising merchants who have invested their millions in iron ships during the past few years, under the feeling of confidence which followed the defeat of the free ship scheme in 1871. It would compel us to depend in an emergency upon the English ship yards, and thus to put labor into the hands and bread into the mouths of English workmen. And it would give English ship owners a possible chance to work off some of their old surplus stock of ships.

SOME PLAIN FACTS.

Here I earnestly call your attention as patriotic citizens to the fact—the truth of which I can vouch for—that there are in New York city to-day, in the hands of brokers and middlemen, printed lists of over 350 English ships for sale, with their speed, tonnage, and draft accurately described. Many of these vessels were built as blockade runners for the Confederate service, and are unfit for use, being such ships as were referred to by Mr. Plimsoll in his speeches before the House of Commons. The real object-of-the men behind this free ship bill is to bring this worn and weather-beaten fleet here under fictitious bills of sale ; run the best of them in opposition to our merchants, breaking down our coast and other trade ; condemn those unfit for use, and sell the old brass, iron, and steel, and thus get rid of the duty on all this material.

The following description, taken from this list, will give you an idea of the fleet. The list is in my possession, and can be seen at any time if you desire to investigate the matter thoroughly:

SCREW STEAMER—3,200 gross, 2,100 tons net register ; built 1853. Barque rig; 3 decks. Accommodation for a large number of passengers. Dimensions 340 x 42 x 34. Compound engines 350 h. p., new 1872. Has done very little work since new engines fitted, and lately been thoroughly overhauled. Would be sold very cheap. [Such a ship might be able to carry a large number of passengers, but how many would care to take their lives in their hands and sail in a bottom 27 years old ? The speed is wisely withheld.]

SCREW STEAMER—2,800 gross, 1,800 tons net register ; built 1863. Large measurement capacity. Dimensions 320 x 43 x 33. Compound engines 500 h. p., new 1873. Consumption 45 tons per day. Speed 13 knots.

SCREW STEAMER—2,700 gross, 1,600 tons net register ; built 1860. Classed 100 A 1. Handsome passenger accommodations. Compound engines 600 h. p., new 1873. Consumption 40 tons. Speed 12 knots.

SCREW STEAMER—2,250 tons gross ; built 1865. Classed 100 A 1. Accommodation for 30 first-class passengers. Compound engines 250 h. p., new 1873. Consumption 24 tons. Speed 10 knots.

These are examples of the 350 ships which England has on hand for sale. No doubt they would all be sold cheap. They certainly ought to be, if sold at all. Estimating these vessels at $100,000 each, their value would be $35,000,000. Our merchants do not want them, for they would be dear at any price, and no prudent man would risk a cargo in them. Would it be fair treatment to those of our own people who have invested their capital to build up our shipping—and in this way given employment to thousands of our own countrymen, and paid taxes toward the support of our government —to allow this fleet to come into our market as proposed? This is a very fine scheme; but once made known, will its concoctors be able longer to get the help of those representatives of our people who have with honest motives and without knowledge of these facts hitherto favored the free ship bill? In presenting his list for 1880, one of the English brokers uses the following significant language:

"When issuing, on 1st July last, a review of the previous half-year, a most discouraging and gloomy account had to be given of the current prices and prospects for shipping, but happily the directly opposite is now the position. * * Should the bill now before the American Congress for removing the restrictions on British-built vessels be passed, extensive purchasers may be expected in our American friends, who, considering the large amount they contribute to the employment of our ocean traffic, ought to occupy a more important position as ship owners."

All will agree to this last statement; but will anybody say why, having already contributed millions upon millions to foreigners for carrying our freights, we should now pay them millions more to support their industries in building for us what we have every resource to build for ourselves? The broker fittingly concludes by saying that "American freights have been the chief support to our shipping during the past year."

SPYING OUT THE LAND.

Is there nothing significant in the fact that there are to-day in the hands of reputable parties in this country letters written by English ship owners to American engineers and other persons having thorough knowledge of the subject, asking for information with regard to our river and coast trade, the draft of water, speed of vessels required, and the best routes on which to place transportation lines? These letters I have seen, and they can be produced if necessary to substantiate any statements made as to a foreign interest in our carrying business. What do they mean? Since none of these English ships can be registered here unless our just and necessary protective Laws are first repealed, do these letters disclose any reason why foreign merchants, ship owners and ship builders should wish for that repeal?

SEEKING CHANCE FOR PROTECTION.

Again, as the owner of a vast fleet, with nearly $600,000,000 invested in it, England has another reason for strong interest in the repeal of our Navigation Laws. In case she became involved in war, which has been hanging over her for years, how could she protect her carriers in their business of transporting her own and the other nations' products? With her great navy of iron clads she could meet her enemy's navy, and could protect her harbors, but not these carriers, and that immense property would inevitably be disastrously affected. But were we to repeal our Navigation Laws, by whitewashed bills of sale, friendly mortgages (like friendly confessed judgments), and other means well known, England could secure for her merchant marine the protection of our flag, and take her fleet back when the war was over, thus retaining her carrying control. It is worthy of notice, that the present free ship bill before Congress, and all the free ship bills hitherto presented, have never put any restrictions on the transfer of these ships back and forth to suit the pleasure of their owners.

Our flag is the only one that can be counted on to remain neutral, since our government alone is able to keep aloof from foreign complications, and does so, in accordance with Washington's advice. How would our people like to have English ships—falsely called American for the time being—unloading under the protection of our flag the products of the English factory in the world's markets, in direct competition with our own ships carrying the products of our own factories? And how, having crushed out our own ship-building interest, could we improve the opportunity of such a war—as England did that of our rebellion—to regain what she thus took from us? If we are ever to be restored to our proper position on the ocean, it must be by means of ships honestly owned by our own people, not by accessions of the character above pointed out. And if we are to hold that position when regained, we must be able within ourselves to sustain it all times and in all circumstances. We need not mind about sustaining anybody else.

Who Wants it Most.

In view of the plain facts here presented, let me ask you, as practical men and citizens concerned in your country's welfare, who appear more deeply interested in the free ship bill : 1. The owners of this $600,000,000 worth of shipping property who want the protection of our flag in case of a European war ; 2. The owners of these 350 ships valued at $35,000,000, who can find no market for them unless our Laws are repealed; or, 3. The American merchants who, if they could buy seaworthy ships at home as cheap as they could buy seaworthy ships in England, could not run them in competition with English owners for the reasons already shown.

Surely if the free ship agitation be honest, it should first and chiefly proceed from our merchants who have invested their millions in ships, and who have given the subject of our carrying trade that thorough investigation which keen men are apt to give the business into which they propose to put their fortunes. What is the fact? I have recently made personal inquiry of the men in this country who within the last ten years have built iron steamships to the amount of many millions of dollars—in fact all that have been built in the country. More than $15,000,000 have been invested in steamships engaged solely in the foreign trade. These men must have closely studied the situation and well understood their business, having to fight against competitors long established and backed by the power of un-limited capital. They put their ships into the foreign trade wherever the foreign competition was not too strong for them, wherever there was the bare chance of getting a hold. But these men and others like them, whose enterprise and capital are devoted to building up our shipping interest on a true and national foundation, do not favor the repeal of our Navigation Laws. They see the real difficulties in the way, which should first be re-moved. They know that the remedy for our present depression in the car-rying trade would not even be touched by a free ship bill, but the disease made worse. Not one of these men ever appears before a Congressional Committee and asks to have the Navigation Laws repealed. Add to this the fact already shown, that under the present discriminations, entirely aside from the Navigation Laws, our own citizens cannot profitably own ships, and is not the conclusion inevitable that the movement for the repeal of those Laws is in a foreign interest?

VIII.

ENGLAND'S PRESENT POSITION.

To show still further England's concern in our shipping and tariff legis-lation, let us look briefly at the present condition of her three great sources of national prosperity.

AGRICULTURE.

Her agricultural interests, as you are aware, are in a most unsatisfactory condition. Her tenant farmers are suffering under the false system of land tenure and from other causes. Labor in agriculture is not less dissatisfied than in other branches, and higher wages or emigration are the horns of the dilemma which her statesmen have to meet. What bread she raises at home is certain to be dearer if the price of labor rises; while in the most favorable circumstances she cannot hope to produce her own bread, and must buy both breadstuffs and meats from us, her great rival in trade. The lack of agricultural resources has always been her weakness, and must remain so. She has nothing to look for in that direction.

MANUFACTURES.

England's great bulwark has been and is the advanced condition of her manufacturing interests. But here also she finds difficulties. One of her largest articles of export, cotton goods, has long suffered from depression without much prospect of change for the better. She is dependent upon us for the raw cotton, and must transport it 3,000 miles to get it to her factories, while our factories have it almost at their doors. She has, moreover, to a great extent lost this market, which she formerly monopolized, both in cotton and other manufactures. Under our wise protective laws and through the reduction in the cost of labor, effected by the use of improved machinery, our manufacturers are not only supplying largely the wants of our own people—almost wholly so in the coarser textile fabrics—but are entering into competition with England in some of her foreign markets. Give us her cheap money ; a few years further time in the development of our resources ; an opportunity to reduce our State, county, city and town taxes, created for improvements made during the past eighteen years at war prices and long a heavy burden on our people ; the balance of trade five years in our favor ; and then, by reason of our many advantages in bread and raw material, we can doubtless meet her in free competition in a great many lines of goods, without either reducing the cost of our labor, or injuring any of our interests.

It appears, therefore, how important it is to England that every possible means be used to prevent us from growing as manufacturers and enlarging our competition in the foreign markets, as well as monopolizing our own. Other nations, moreover, are following our example, at least in trying to supply their own wants ; and to that extent they also reduce England's control of her former markets, and her greatness as the chief manufacturing nation of the world.

KIND ADVICE FROM ABROAD.

In this connection, it is significant to notice that English writers upon our tariff legislation frequently express the opinion that the people of this country have made a mistake in undertaking any branch of manufacturing industry, and that we would have been much more prosperous had we confined our attention mainly to agriculture. This very kind and considerate view, and its unselfishness, are set forth in this wise in a series of articles on "The resources of Foreign Countries," by A. J. Wilson, an English writer :

"There is no use in denying the plain fact that the States have succeeded by their high tariff policy in diverting a considerable part of the industrial energies of the community from the pursuits natural to, and most profitable in, a new country, to the highly artificial, and, for America, mostly very expensive industries of long settled and civilized nations. Were the sheltering tariff swept away, it is very questionable if any, save a few special manufactures of certain kinds of tools, machinery, railway cars and fancy goods, and a few of the cruder manufactures, could maintain their ground."

Entertaining such notions as this about our ability to sustain our manufactures under free trade, and realizing that at present we are making very strong competition against their own goods, is it strange that the English think it would be a very fine thing for us to repeal our protective tariff and

Navigation Laws? As to our being a civilized country, we can afford to let that pass, considering the show we made to the world in our own Centennial, and later at Paris. No doubt England, finding her own manufactures depressed and her markets curtailed by our competition, earnestly believes it would be vastly better for us to devote all our attention to raising bread and cotton. But probably our own people have some opinion and voice upon that subject.

A WAIL FROM FREE TRADERS.

The following extract from the Annual Cotton Circular of a Liverpool House is also interesting reading :

"Then this country has suffered very severely of late years from the increasing stringency of foreign tariffs. There has been a growing tendency evinced in most countries to protect their own industries, and in every such case we are the chief sufferers, for we live as already said, by exchanging our manufactures for the necessaries of life. The United States was at one time a large customer for our iron ware and textile fabrics, but the hostile tariff she has enforced since the civil war has nearly driven us out of her markets, and has built up a vast system of manufactures which completely supplies her own wants, and leaves something to spare for competition with us in foreign markets. The free traders of this country console themselves by thinking that she is the chief sufferer, but whether this be so or not (which is very doubtful) the fact remains that her markets are almost lost to us, and we, on the other hand, are constantly more dependent upon her for food and raw material. For this we have no means of paying except by money or bonds, or indirectly by our credits with China, Brazil, and other countries, from which America imports tea, sugar, &c. Our colonies all follow in the wake of the United States, and do their best to stimulate their own manufactures by closing their markets against ours.

"The countries of Europe, on the other hand, aided by our best machinery and skilled labor, which we have largely supplied, are becoming more and more independent of us, and in those cases where we can still undersell them are raising their tariff or threatening to do so, and the last stroke in that direction is exhibited by Prince Bismarck's letter to the enlightened German nation, wherein he boldly advocates returning to protection ; and we are given to believe that legislation in that sense is pretty sure to take place. The existence of our trade is therefore restricted to India and China, and such weak semi-civilized countries as are in some degree under our control, and no doubt in these "open markets" we can still hold our ground ; but, unluckily, terrible famines have wasted the resources of both India and China of late years, and we are still feeling their effects."

Here it is admitted that England is partly dependent upon the money which we send to China and Brazil. Suppose, then, we were to pay for our imports from those countries in goods, the very thing which our merchants are aiming to do, what result would that have upon England's treasury? This partly explains the desperate effort making to break down the new American line to Brazil, and the bitter opposition encountered from the first to the proposed encouragement by our government of that essential means of opening to our merchants and people this great South American market, now almost wholly in England's control.

ALMOST TOO GOOD NEWS.

Further, the following cablegram recently sent from London to the New York *Tribune*, is well worthy attention, as showing the anxiety with which

England's manufacturers are waiting legislation by our Congress, and what they hope to gain thereby :

"The announcement of the introduction of bills in the House of Representatives at Washington, proposing the reduction of iron duties, rouses the liveliest hopes among British manufacturers. Leading journals in the iron districts hail the prospect of once more arresting the present development in American iron and steel manufacture. *The Newcastle Chronicle* declares it has reason to entertain great hopes of the success of these measures. It considers the free admission of iron ore as intended to secure a Canadian supply, but it would result in increasing the ore imports from England and Spain. If free ore were secured free coal would necessarily follow, with a general increase in English exports. *The Chronicle* declares these measures to be the most important news cabled for a long time. *The Economist* says their effects upon the English iron trade would be enormous. It fears lest the proposals may be too favorable to English trade to have any chance of adoption."

The depression in England's manufactures thus apparent and admitted, and our growing competition recognized, it will be seen that with respect to this great interest she has anything but a hopeful outlook.

THE CARRYING TRADE.

It is in the third great interest—the carrying trade on the ocean—that England is prosperous. Upon this depends her power among the nations, and to maintain this in the proud position it now occupies, all the vigilance and shrewdness and energy of her statesmen and people, all the influence of her vast capital will be exerted. From whatever direction the competition comes, it will be promptly met if a policy can be devised to meet it. I beg you to carefully consider the magnitude of England's interest in the carrying trade, as shown in the following chapter, the figures in which are authentic.

IX.

A VAST INTEREST.

According to the Bureau of *Veritas* of France, the sea-going tonnage of the world in January, 1879, was as follows :

THE WORLD'S TONNAGE.

Character.	No.	Tonnage.	Estimated Value.*
Sailing Vessels..........	49,524	14,317,430	$572,692,000
Steam Vessels..........	5,462	5,524,797	552,479,700
Total.................	54,986	19,842,227	$1,125,171,700

*The estimated value is at $40 per ton for sailing vessels $100 per ton for steam | a low estimate.

This tonnage is divided among the four principal nations as follows :

*ENGLAND'S TONNAGE.**

Character.	No.	Tonnage.	Estimated Value.*
Sailing Vessels	18,394	5,696,018	$227,840,720
Steam Vessels...........	3,216	3,465,187	346,518,700
Total..............	21,610	9,161,205	$574,359,420

*It must be borne in mind that one ton of steam is equal in capacity to three tons of sail. England's excess of steam tonnage in proportion to that of other nations will therefore be an important factor in considering her capacity.

*AMERICA'S TONNAGE.**

Character.	. No.	Tonnage.	Estimated Value.*
Sailing Ships............	6,050	2,075,832	$83,033,280
Steam Ships.............	516	609,111	60,911,100
Total..............	6,566	2,684,943	$143,944,380

*This includes the sea-going coasters.

GERMANY'S TONNAGE.

Character.	No.	Tonnage.	Estimated Value.
Sailing Ships...........	3,201	914,674	$36,586,960
Steam Ships.............	220	253,667	25,366,700
Total..............	3,421	1,168,341	$61,953,660

FRANCE'S TONNAGE.

Character.	No.	Tonnage.	Estimated Value.
Sailing Ships...........	2,972	595,933	$23,857,320
Steam Ships.............	275	335,219	33,521,900
Total..............	3,247	931,152	$57,359,220

The balance of the tonnage is scattered among nearly a score of nations, and is not important to consider.

A RESUME.

The shipping interest of the four principal nations is thus shown to be :

Nations.	No.	Tonnage.	Value.
Total Ships.............	54,986	19,842,227	$1,125,171,700
England................	21,610	9,161,205	574,359,420
America................	6,566	2,684,943	143,944,380
Germany...............	3,421	1,168,341	61,953,660
France.................	3,247	931,152	57,359,220

A comparative table of steam tonnage gives these results:

Nations.	No.	Tonnage.	Value.
Total Steam Ships	5,462	5,524,797	$552,479,700
England	3,216	3,465,187	346,518,700
America	516	609,111	60,911,100
Germany	220	253,667	25,366,700
France	275	335,219	33,521,900

INTERESTING CONCLUSIONS.

From this it will be seen that our total tonnage is about one-fourth that of England, and the tonnage of Germany and France about one-tenth. Of steam tonnage, however, which is three times greater in capacity than sail, as stated, we have only one-sixth as much as England, and France and Germany together about equal our tonnage, leaving England more than three times as much steam tonnage as the other three great nations combined. Hence it appears that England's actual carrying capacity is equal to three-fifths of the total tonnage of the world. This is surely a proud position for one nation to occupy. Three-fifths of the world's carrying trade conducted under her flag, and the other nations, with the single exception of our own, almost wholly dependent upon her ship yards for their ships. And though we are not yet dependent upon those ship yards, the advocates of the free ship bill are doing all they can to make us so; while the policy we have pursued has already made us dependent upon her for the carrying of the great bulk of our products to the markets of the world.

In regard to the $143,944,380 worth of ships owned by our citizens, it should be said that the greater part of this property has been created since 1860, very much of it at war prices and at a time when our dollar was worth only seventy-five cents. Its construction not only gave employment to thousands of our people, but upon this property heavy taxation was paid for the benefit of the government. A large amount of capital is now invested to keep this fleet in repairs, and in that work and in manning these ships many thousands more of American citizens are kept employed. Certainly the men who had the enterprise to invest this $143,944,380 under so unfavorable circumstances are entitled to some consideration. It could scarcely be considered fair treatment now that the outlook is a little brighter, to allow foreign owners to send over here ships built by foreign labor, paying taxes to a foreign government, to break up their business.

ADVANTAGES OF SHIP BUILDING.

England knows that we are her only formidable rival, because ours is the only country that possesses the natural resources to build ships. That none but a ship-building nation—meaning by that a nation having within its territory the required natural resources—can compete successfully with her in the carrying trade has been sufficiently proved by the experiments of France and Germany. Those countries have every advantage possessed by England except the natural resources ; and as to that they have free trade with England in iron and coal. France last year bought 10,000,000 tons of coal, principally from England ; and England's coal and iron are nearer to the rivers in France where ships could be built than the French mines are, the freight from the English mine to the French ship yard being only one-half that from the French mine to the same yard. Not only have France and Germany free trade in the raw materials, if they want to try building ships at home, but they have as well what the free ship men claim would be so superlative an advantage to us, the liberty and privilege to buy their ships from England free of duty. Their advantages over us, therefore, with the sole exception of natural resources, are numerous. They have as cheap labor as England has with which to build and run the ship ; they have as cheap capital to own it, and but one-tenth of the taxation on that capital which the American company must bear; and besides all this, they have liberal legislation which awards them generous contracts for their ocean mail service. Yet with so much in their favor, they have only one-tenth the tonnage that England has. Will the free ship advocate tell us why this is so ?

The fact remains, and England well understands it—so well that, if by any means she could get our ship yards closed, she would feel secure in her position, not only as the world's carrier, but as the great workshop to which all the nations must come for ships and tools, whether for their navies or their commerce. If France and Germany lose the ships they buy from England, must they not buy new ones, and keep on buying? Must they not drain themselves constantly of the large amounts necessary to keep a fleet in repairs ? And if we begin to buy our ships of England, will we not be in the same position of constant outlay and dependence ?

A PERTINENT QUESTION.

Here it may well be asked, could England afford to own this $574,000,000 worth of ships if she had to buy them from another country? To keep this fleet up, not counting the annual cost of repairs, to build the new ships required to replace those lost and worn out, and to meet the demands of increased trade, costs not less than ten per cent. of the whole value, or $57,000,000 a year. Could any nation expect to maintain such a mighty interest as this if, instead of being able to build its ships and tools, it was dependent upon another nation for them, and must annually send millions upon millions of gold to support the working people of that other nation ? Could we afford to own such a fleet if we had bought it abroad ? Could we pay out annually the millions necessary to keep it in repair? What effect would the taking of that great amount of labor out of our market have upon it, and what effect in comparison upon the market which we favored with it ? These questions are of deep importance. England's statesmen, seeing how they are thrown back on this interest, closely study all possible chances whence competition may arise, and use every means open to them to prevent that competition.

And you well know the mighty influence which vast capital can command over public opinion, and the many ways in which it works to accomplish its purposes. Its influence extends to even the most honest of our people, and discloses itself constantly in one way or another. Much of the sentiment manifested in favor of free ships is manufactured by its mouthpieces, and spreads easily among those who have not time or opportunity to thoroughly investigate the subject. The influence of this enormous capital, moreover, has had nothing to contend with here save the efforts of a handful of our enterprising ship owners and builders, who for their endeavors to build up the industrial interests of their own country and to regain for her the place once held, and now rightfully hers, on the ocean, have met only with opprobrium and denunciation by a portion of our press as monopolists and subsidy grabbers. Even many of England's liberal statesmen, who are considered true friends of America, have found it in their way to deride us because we do not repeal our Navigation Laws and let England build our ships. They would prefer that we should do more of the world's business and they less, provided we would make them our carriers and builders. But whatever good feeling they may have for us, and whatever persuasive arguments they may use, I am satisfied that they are always for England first and for us afterwards.

X.

WISE ENCOURAGEMENT.

A moment as to national legislative encouragement. The English steam tonnage receives about $4,000,000 a year for mail service. From 1867 to 1876 England paid in this way $52,138,837. France, who followed England's example in 1858, and made liberal mail contracts with lines to the Levant, to New York, to Brazil, the West Indies, and Mexico, is to-day paying $4,800,000 annually on a steam tonnage of $33,521,900. She has maintained this policy through all political vicissitudes, including the revolution from Empire to Republic. Her statesmen have recognized the fact that by such a course alone could France hold place on the seas, or in the markets of the world where she must put her surplus products. Brazil pays $1,500,000 to the steamship lines which carry her mails, a service which has been almost monopolized by the English. Germany pays on a tonnage of $25,366,700. Austria, Italy, and Spain pay for mail service, to extend their trade in the Mediterranean and also on the Atlantic : while even China and Japan are not behind the Western Nations in this policy.

But the United States, with more surplus products to carry on the sea than any other nation, with pressing need to open up new markets, and hence to establish rapid means of communication between its merchants and the merchants of the countries whose markets are naturally open to it, alone refuses to pay its own steamship lines fair compensation for the service they perform. At the same time, it compels them to carry the mails for postage money merely. This postage it also pays to the various foreign lines running to this country. In postage, the United States in 1876 paid $182,863.30 to help support foreign carriers. From nearly all the countries to which they run, except ours, these foreign lines also receive contracts. In this way other governments have met England in her liberal policy, until where she pays $4,000,000, she gets at least $2,000,000 from them in return, in contracts for mail service, besides the vast profits in freight and trade. Where such return encouragement is refused, England is perfectly willing to appropriate whatever amount is necessary, knowing from fortunate experience how many times over she will get back its value.

EASY GENEROSITY.

Thus England and France have always been so kind to us that they would pay to their own steamship lines sufficiently large compensation, in the form of mail contracts for a term of years, to enable them to carry our mails for the postage only. What has been the result gained by this truly generous policy? England to-day has the carrying business of our country, and drains us from $60,000,000 to $90,000,000 of gold every year in freight and passage money. The American economist of a certain class will exclaim on the extravagance of paying even so much as the postage amounts to for the carrying of our mails; while at the idea of a bill authorizing the payment to an American line of a small sum for ocean mail service he is, or pretends to be, horror stricken. In his specious plea for economy, he never considers that England by paying $4,000,000 a year in mail contracts draws to her island over fifty times that number of millions of dollars, and maintains herself as the banker of the world. He does not appear to reflect that to pay out $4,000,000 and get $70,000,000 in return is a profitable investment for a country's interests. Yet he talks about economy, and the money he wishes to save his country, and is unaware how completely he is arguing our rival's cause.

It is often stated that England does not now pay by contract for mail service on the North Atlantic. This is not true, a new contract having been made in 1877. But what would it signify if it were true? Simply that England had by encouragement built up her steamship lines on the North Atlantic to that degree that they no longer needed help. England pays her mail contracts in the same spirit with which in olden time she used to send out her war ships and cannonade the merchant marine of her rivals from the seas. When she had crushed out one rival and it was no longer necessary to keep her fleet in pursuit of that enemy, she simply transferred her forces to the next point where they were needed. She has pursued the same course with regard to her mail contracts. Her policy, as shown in correspondence between the Postmaster-General and Lords Commissioners of the Treasury, is to render succor to such lines as meet with opposition of foreign tonnage, and to such as meet with unavoidable reverses ; leaving those lines having an uninterrupted sway to take care of themselves. Is not that certainly a shrewd and statesmanlike policy? It has built up for England the most powerful merchant marine in the world. This policy she will continue, and there is no doubt that, in case lively competition were threatened, she would be ready to pay us for the privilege of carrying our mails, if we would discontinue efforts at competition and allow her to keep on carrying our freight and passengers also. She certainly could well afford to do it.

OUR PROBLEM.

How can American lines, without pay for the service they render government, compete with rival lines thus established and supported? Some will utter the old cry, "Repeal the Navigation Laws, and let us go abroad and buy ships cheap, and we can compete." That is a good illustration of the blind way in which every evil is put upon the poor Navigation Laws, and every remedy found in their repeal, without regard to common sense. But let us see. The only market where we can buy ships is England. Is it not likely that she builds ships as cheap for her own merchants as she would for us? Yet she has paid and now pays her own companies for their mail service. France, Germany, and Spain buy their ships from England. But these nations, too, pay for ocean service. How then would our companies, simply by buying ships from England, at whatever price, be able to run them at the high speed required of such steamers, without like encouragement? Why does not the advocate of free ships apply himself to explaining some of these questions, instead of forever seeing but the one thing, which is not the real thing at all?

SINGULAR UNSELFISHNESS.

The free ship advocate who works for foreign interests is always opposed to the payment by the government of a just compensation for carrying the mails. Though he knows that other governments have paid and are to-day paying large amounts for this service, he wants it done for nothing. He, of course, is not opposed to liberal payment for the coasting trade, because he knows that will not interfere with foreign interests. Sometimes, indeed, he becomes quite generous and will admit the justice of the demand for mail contracts; but he will always make it conditional that ships built by foreign labor and to be paid for by American capital— the taxes on whose construction goes to the English government—shall receive here the same recognition and encouragement as ships built at home,

by American labor, paying construction and other taxes to our government.
Could England's interest be better represented in her own Parliament than
they are here by this sort of free ship advocate? He is the first man I ever
knew who was opposed to receiving pay for services rendered. He is truly
a marvellously self-denying man, where America's interests are concerned.
Perhaps he can afford to be, since he has no idea of starting a steamship
line under any conditions.

XI.

OUR NAVY.

There are many urgent reasons why we should not repeal the Navigation Laws, but should on the contrary employ every means under them to
build up a powerful and effective merchant marine. The United States
occupy a peculiar position geographically. We have 15,000 miles of coast
to protect, with an immense territory bordering on two oceans. Besides
this comes the call for protection from our shipping engaged in both the
coasting and the foreign trade. How shall this protection be secured? Is
it not manifestly the wise course to secure it in that way that will cost least
in time of peace?

Before the introduction of steam into the navies of the world, and
when all were alike dependent upon wind and sail, no nation had a more
efficient navy or a more thoroughly equipped merchant marine than the
United States. Since that introduction, though to America belongs the
honor of first crossing the Atlantic by steam, we have lost position until
we are far inferior to what we should be on the ocean. It would seem that
our government has had no fixed or settled policy with regard either to our
navy or merchant marine, while foreign nations have paid very close attention to both. Since steam was introduced to any extent into naval service,
our government certainly has never had any carefully devised and well
defined plan for the construction of a navy. Whenever anything has been
done in that line, it has been done after the emergency came upon us, when
all was confusion and alarm, and the work of a year must be crowded into
three months. That old cry, "We must do the best we can in the crisis
which is upon us!" has often been heard by our people; and when the
crisis was over—some sort of means to weather it having been patched up
at great expense—the same listless, careless policy was again pursued.
This course has cost the country millions of dollars, many times more millions than were needed, if expended under the right policy, to have made
and maintained for us the most powerful and profitable marine afloat. And
after all the expense, what was there to show for it?

If ever a nation paid the penalty for neglect to provide the means of defence and protection in emergency, this nation paid it in blood and treasure during our civil war. The want of a navy to effectively blockade the Southern coast operated against us in two ways: 1. The English blockade runners were enabled to supply the South with resources to carry on the war, and thus did more to prolong it than most people are aware; and, 2. Our merchant steam marine was absorbed in the endeavor to look after them; and in other service for government many hundred thousands of tons of our merchant shipping were required. It may be said here that before our war, no nation criticised and condemned the cause of it more than England; but when the war came, no other nation so interfered with our blockade and aided to keep up the strife. Had we been prepared with a powerful fleet, as I have shown in a preceding paragraph, the war might have been ended within a year. For want of that preparation we suffered everything short of the destruction of our government.

INSTRUCTIVE COMPARISONS.

In considering our navy let us look first at the position of affairs in 1858. This period is selected to show the condition of our steam tonnage when the war broke out, and the terrible mistake made years before in not encouraging our merchants to put fast steamships on the ocean, to meet those built by England; also to show that to a great extent our government at that time is to be charged with the destruction of our carrying trade, through its neglect either to pursue a policy that would render our marine self-protective, or to provide for its protection by government means.

In 1858 our sea-going tonnage was as follows :

Sailing ships, in foreign and coasting trade..	5,220,000 tons.
Steam vessels...............................	71,000 tons.
Total fleet of merchant marine...........	5,291,000 tons.

This steam tonnage was composed principally of wooden side-wheel steamers. There was no ocean-going iron steamer then in the country. To protect this fleet we had the following steam navy:

Fourteen wooden steamers: Eight of them of 3,000 tons each, speed not to exceed nine knots; three side-wheel steamers, two tenders, one dispatch boat, of about 558 tons each. This fleet was scattered in the foreign stations and along our coast, and composed our formidable armament on the sea.

AN AMUSING ILLUSTRATION.

To illustrate the effectiveness of this navy, and the kind of protection it was able to give our carrying trade, an incident that occurred in the latter part of President Buchanan's administration will serve. You may recall that in 1858-9 some difficulty arose between our government and Uruguay in relation to the firing upon one of our ships and the killing of an American citizen. Some small sum was asked in reparation, and refused. To maintain our citizens in their rights something had to be done, and it was decided to send a squadron down to obtain satisfaction. To make up a fleet government bought or chartered in New York five ships of the old Cromwell line, and fitted them out for a navy. These vessels could not carry their own coal, and had a speed of not over nine knots. It took them seventy-two days to go down to Uruguay. By the time they arrived the matter had been settled, and all they had to do was to take seventy-two days more and come back. On the return they were ordered to keep close together, so that in case of accident—it not being at all certain but one or two of them would go to the bottom at any moment—they might help each other out.

That was the way in which we constructed a naval squadron, at a time when England alone had in her merchant marine 152 iron steamships, many of them able to make fourteen and fifteen knots, all of them to run away from and around any ships which we owned. Such a disclosure of naval weakness had the effect to stir up our government to some show of activity in regard to the navy, but little came of it in actual results. From this incident you can judge what kind of a steam navy we had prior to the rebellion, and the position we occupied before the nations as a naval power.

Working Under Difficulties.

When the war broke out, the first thing to be done was to remedy so far as possible our naval weakness. As we could command plenty of soldiers, so we could of sailors and seamen. The trouble was not that of getting good naval officers and men—for ours had no superiors in bravery and fighting qualities; it was to get ships suitable for them to fight in. It was not so easy at short notice to raise the army of skilled workmen and the facilities required to build for us a steam navy. The other great nations were then experimenting with steam in their navies. We had no time to join in the experiments, nor to deliberate over plans. The crisis was upon us, everything was in confusion, and an immediate demand was made upon the resources of the country. Nobody thought of economy in that hour. Any and every sort of ship that could be bought or built for a special service was secured, without regard either to cost or future fitness for our navy. In this way large amounts of money were spent without profitable results. During that same period England and France expended equally large sums in their experiments upon iron steam war ships, though they were at peace. While they, too, had little to show for their outlay, since it is but recently that the more formidable war ships have been built—they had laid the foundation for their present powerful navies.

As a consequence of being taken unawares, many of the ships which we succeeded in building were of green timber and rotted almost as soon as built; and but for the aid given by the private workshops and the enterprise of our builders, we should have been unable to obtain any helpful service on the sea. A single incident will show what these private facilities, which

it is now proposed to close by the free ship bill, did for us in our crisis.
When Washington and our fleet were alike threatened with capture and destruction, what saved us ? A little Monitor, which had been built in a private yard in New York in ninety days, conquered the vaunted Merrimac,
rescued our imperiled fleet and capitol, and threw the enemy into conster-
nation.

The fact is noteworthy that the vast amounts paid by our government
for naval purposes during the war were not spent to crush out the Southern
Confederacy so much as to meet the emergency threatening us from other
quarters. The great draft upon our resources made the cost of everything
connected with the navy much heavier than it would have been had the
things been bought at the proper time. What cost us millions in the war
could have been obtained for thousands before the war came ; and that they
were not so obtained must be charged upon a policy of inactivity not unlike
that which we have since pursued. What have we done to improve our
position? How are we better prepared for an emergency to-day than we
were in 1860, in comparison with the great naval advances made by other
nations?

Further Comparisons.

Coming to the present time, we find that our sea-going tonnage in the
foreign and coasting trade was, in 1879, as follows :

Sailing ships, 6,650, with tonnage of..............	2,075,832
Steam ships, 516, with tonnage of.................	609,111
Total tonnage......................................	2,684,943

or a total tonnage of about one-half that owned by us in 1860. To protect
this fleet, which is valued at $143,944,380, what navy have we ? The fol-
lowing list will show :

Tugs for harbor use..............................	25
Sailing vessels, of little use...........................	13
Monitors, for harbor defence..........................	16
Steamers...	54
Total..	110

Of this fleet of fifty-four steamers, one is new—the Trenton—with speed
of fourteen knots ; three have speed of thirteen to fourteen knots ; four a
speed of about twelve. The balance consists of slow steamers, large consumers of coal, furnished with a style of machinery long since discarded by
the merchant marine of the world, and many of them old and unserviceable.
The fact is, after you take out the iron clad monitors and a few ships
recently built which have speed, there are not twenty-five ships in this
whole fleet worth preserving.

Our Navy Yards.

To take care of this navy, which certainly is not strong enough to take
care of itself, we have eight navy yards, with grounds, buildings, machinery and equipments valued at $100,000,000. It is wise to have these as a

reserve power, and they should unquestionably not only be maintained in the highest state of efficiency, but be increased if we are to follow our present system of discouragement to ship-building, and continue to threaten our owners and builders with the free ship bill. But to-day there are only some 3,000 men employed in these yards, which have capacity to employ 60,000 mechanics. It has not been our policy to keep a large working force in them while at peace, nor to make them schools for educating mechanics. On the contrary, to save the millions annually which that would cost, the government has depended upon the private ship yards and workshops (to build up and maintain which it never paid a dollar) to fill its navy yards with skilled mechanics and engineers in case of emergency. Thus in our war the private ship yards and workshops furnished an army of 60,000 skilled mechanics, without which the navy yards would have been worthless. It is undoubtedly true economy to pursue this policy, and it proves the importance in every regard of developing and maintaining these great private schools of skilled labor. When they sent out that army of trained mechanics, did they not thereby prove in a manner that should not soon be forgotten by our people, the value of those Navigation and other laws which had protected and kept them in existence? Suppose these private sources had been closed prior to 1860, by the same means now being used in Congress to close them. In that case, where could our government have found the necessary skilled labor—in the yards on the Clyde? If we are going to shut up these practical schools, through penny wise notions, we must provide some other means to furnish us with trained labor of a high class. If we lose this resource we cannot so easily replace it as we could an army of sailors. Who can estimate the value of those industries, which in peace cost the nation nothing, but in her hour of necessity give to her service a skilled force that no amount of money could procure at short notice, a force that must be educated year by year? The free ship advocate, who has much to say about the loss of American sailors, has nothing to say about the loss of American mechanics, though that loss would be incomparably greater.

PRACTICAL SCHOOLS.

Strangely enough, while the necessity of having a vast force of skilled mechanics and engineers in the country is recognized by all, free trade or otherwise, the equal necessity of maintaining the training places for this force is lost sight of. How shall we expect to have the one without the other—the effect without the cause? This would be like admitting the need of general education, passing a compulsory law to secure it, and then abolishing the public schools.

Is it said that the free ship bill will not close the workshops outside of the ship yards? I answer that the same principle applied to our other industries will close them all; and surely the ship builder is not to be singled out as the solitary victim of a theory. Besides, more than thirty branches of industry enter into ship building. I have spoken of the 60,000 skilled mechanics and engineers furnished to the government at a time when their aid was incalculably valuable. Why, nearly all our present corps of naval engineers are graduates from our private workshops; men who are not only engineers but practical mechanics; who can take an engine apart and put it up again, or take the raw material and work it into its parts. These men possess a knowledge of their profession which cannot be learned in any other way than in the workshop, cannot be taught in the academies, however efficient, thorough, and necessary these may be and unquestionably are. Our ship yards are the special practical schools in which are trained the mechanics upon whom the government relies mainly to fill its navy yards in case of necessity. What kind of interest in his country can the man have who says these should be closed, simply because we can buy ships a little cheaper in a foreign country—and can buy them cheaper, why? Because, as shown elsewhere, we refuse as a nation to degrade labor to the position it occupies in Europe. Go to one of our great workshops at noon-time, and look into the dinner-kettles of the workmen. That will give you a good part of the whole story in a small compass. While it is undoubtedly wise to maintain our system of education at the naval academy and polytechnic schools, the engineer on the ship should be a practical mechanic, not only able to start and stop his engine, but to repair it if necessary. More particularly are such engineers needed on our war ships, which are often for months out of port, and reach ports where superior mechanics cannot be found. This point may be illustrated by an incident of our war, much talked of at the time. A train filled with troops was hastening toward an important point, when the engine broke down and there was a stop. There were plenty of men on the train who understood the principles of steam, and numbers who could run an engine. But the demand was for a practical mechanic who could repair the engine. Call was made, and a Boston machinist responded from the ranks, soon put the engine in order, and the train went on. Such a circumstance might happen at a moment of the very greatest importance on a war ship.

It should be said here that our want with respect to a navy has never been that of trained, brave, and efficient men, whether in command or in the ranks. In our navy, indeed, we have men of talent in their profession equal if not superior to any who can be named in the naval forces of other nations. As to the practical knowledge and training of our naval corps I have already spoken. That we should retain a naval force of this high character with so inadequate a fleet is a great credit to our nationality. In case of a naval war to-morrow, we could furnish some of the best material in men in the world. But to send these men into a fight in any war ships we possess would be like the fatal sending of the six hundred into the charge at Balaklava.

Something Like a Navy.

Let us look, now, at England's navy, and see whether her course has been like ours, or whether she has at once built up her commerce and a navy to protect it. We find that her policy in relation to the shipping interest, naval or merchant, has been all of a piece, framed to develop her resources and increase her greatness. In time of peace she constructed her naval ships principally in the private ship yards, maintaining the royal docks as a great reserve to be used in war. Other nations have followed this example by having their navies built in the private yards, with general satisfaction.

The English navy last year consisted of 463 ships of war—principally modern iron steamships, of great speed and effectiveness. There is no ship in our navy to compare with them. In March, 1879, a resolution was passed in Parliament asking the amount of money expended in naval construction and repairs from 1874 to 1879—a period of peace. The following figures are taken from the official reply :

The amount expended in construction and repairs, not including supplies, was £14,296,710 ; seventy-five to eighty per cent. of which was paid to the private ship yards. At that time the English currency was on a gold basis, and its purchasing capacity in her cheap labor market was fully one-third cheaper than ours, which was below par ; so that had this work been done in the United States it would have cost her a third more—£19,062,280, or $95,311,400.

In addition to this England pays annually $4,000,000 for postal service (the mail steamers becoming under their contracts a part of her navy), making $20,000,000 more in the five years. The repairs should have been a small portion of the expenditure, as her navy is mostly new and in good order. Not content even with such a showing, England is now constructing in her private ship yards twenty-five new steel corvettes, of greatest speed ; and doing all this in time of peace, in accordance with her policy of always being prepared ; knowing also that by doing it at such time she gets better work at less expense.

In the same period, from 1874 to 1879, I believe that our government did not appropriate for naval construction and repairs more than $12,000,000; while we did not pay to lines in the foreign trade a dollar as against the millions which England appropriated for mail service on the ocean.

Supplemental Strength.

Yet with all her naval power, England does not depend alone upon her navy. Recognizing the value of having within call a great fleet of fast ships, capable of being easily transferred into cruisers and war ships for the defense of her own and the destruction of an enemy's commerce, she has in every way encouraged the construction of such a fleet. In place of the government building and owning them, inducements were held out to her merchants to construct and employ them. By this means, at small expense she has secured the necessary and desired supplement to her navy, until to-day she has in her merchant marine 3,500 iron screw steamers, at least one hundred of them having a speed of from fourteen to fifteen knots. As many of these as are required can be quickly transformed into cruisers, blockade runners, or war ships. No ship in her navy or in any other can overhaul one of them when once at sea.

With our own bitter experience to warn us, and despite the plain need to have a merchant fleet similar to that of England, our government has never done anything to help build up such a means of defence for us. In view of the great naval improvements made by foreign nations, is it just to the men who have invested their millions of capital in ships to give them no guarantee for the protection of their property save that of such a navy as ours? Should war arise now or at any time, our merchants must tie what ships they own to the docks, unless the ships could defend themselves. That was what they had to do in our rebellion, their ships having no protection; yet they had to pay taxes on them all the same. If it was necessary for England to have such a supplemental fleet, is it not much more necessary for us? If so, shall our government build and own them, at great and entirely needless expense, or would it be wiser to adopt the policy of other nations, and encourage our merchants to build and own them? Can there be any question as to the advantage it would be to this country to have such a fleet of swift iron screw steamships, in time of peace developing the country's wealth by opening up new markets and carrying to them our surplus products, collecting from foreign countries the outbound freights in gold, and from our own merchants the inbound freights on foreign goods, both freights to be spent at home; and in time of war or sudden emergency, ready with trained American seamen and engineers to defend the nation at its call? Such a fleet might well be named the Militia of the Sea. With it, and the power to create it at home of unquestionable nationality, we should possess a mighty safeguard, worth a million-fold whatever trifling expense it might be to the country.

Genuine Economy.

Much is said in Congress and among our people about economy. Will some of our statesmen investigate this question of navy yards, private ship yards, and the free ship bill with express reference to the question of true national economy? To start with it may safely be said that under a policy calculated to build up a powerful iron merchant marine we should not need to maintain all of our eight navy yards.

Look at this point a moment. England builds three-fourths of her naval ships, and nearly all of her marine engines in her private yards. How much she saves by this can easily be seen; besides in this way giving other nations such confidence in her private ship yards that they also get many of their naval ships built there, and send millions of money into her treasury.

With 463 war ships and twenty-five more building, England has four navy
yards, or about 114 ships to a yard. France, with a navy of 342 ships, has
five yards, or about sixty-eight ships to a yard. But the United States,
having 110 ships all told, and in reality fifty-six ships which could be used
effectively, has eight yards, or one navy yard to every seven ships. The
comparison is easily made. Under her policy of patronizing the private
yards and sustaining her merchant marine, England dispenses with the
need of a large number of navy yards; and if the figures could be ascer-
tained, I doubt not they would show that in this way she saves far more
than she expends to encourage her merchant marine.

<h2 style="text-align:center">THE MOST PROFITABLE NAVY.</h2>

With our harbors well protected, we do not need to maintain a great and
expensive navy like the navies of the European nations. But what naval
ships we do have should be of the most improved and effective kind. In
obtaining these ships we have some great advantages. The European governments spent millions in experimenting before they got the war ship they
wanted. We can profit by their mistakes, having the result of their costly
labors to guide us. Experience has proved that in modern naval warfare
the one thing to be desired is speed. This gives a wonderful advantage
either in attack or retreat. To attain this a million of dollars might be
spent in the construction of a war ship, and, though the maximum of speed
might not be required more than five times in the ship's life, for the honor
of the nation it might be of more importance to us in a crisis than ten times
the whole cost. While cruising the extra speed not be employed, and in
consumption of coal and otherwise the fast ship would be no more expensive than a slower one. One iron clad of great speed, it is well to bear in
mind, is in nearly all cases more effective than three slow ships of the same
size and equipment. Our naval officers will need no proof of the advantages of speed. Twenty such ships, capable to meet the naval fleet of an
enemy, would be a vastly superior power to the whole of our present navy.
Yet the ships we have now cost as much in repairs as the others would,
while as to efficiency there could be no comparison. Such a fleet, with the
addition of a number of our present naval vessels, would give us a powerful
navy to defend our coast and commerce. These swift iron clads need not
be built more rapidly than five each year; and being constructed in time of
peace they could be both cheaper and better built than in circumstances
requiring great haste.

Supplement this by a policy that would induce our capitalists to build
and maintain a fleet of forty or fifty very fast merchant steamships such as
I have mentioned previously, and we should have sufficient protection on
land and sea. This auxiliary fleet would involve no expense to the government except the small sum appropriated for postal service—a trifle in comparison with the beneficial returns. If these ships were built for the navy,
when they were lying idle it would cost more to take care of them than to

give such encouragement as would enable our merchants to own them. Besides, in the latter case they would be paying taxes to the government and be productive property, while in the former they would neither pay nor earn anything. With such a navy we should occupy a different position on the ocean from that which is ours to-day, and one far more in harmony with our national greatness.

XII.

AMERICAN PROGRESS.

In this connection permit me to call your attention to the progress we have made in iron ship building during the past seven years, in the face of the difficulties which I have pointed out. In four iron ship yards on the Delaware, from 1872 to 1879, there were built 76 iron screw steamships, of 152,088 tons, giving employment to a crew of 3,986 men. All these vessels are sea-going, and equal in strength, speed and finish to any of their class in the world. All are excellent cruisers, having great speed, and easily adaptable to naval uses. With the number of men employed they make in themselves a respectable navy. Of these vessels 25 have a speed of fourteen knots; 20 a speed of thirteen knots; and the balance a speed of twelve knots. In tonnage they range from 5,300 to 2,000 tons. At the same time there have been built in this country 25 ocean-going wooden screw steamers, with a tonnage of 27,563, most of these steamers having a speed of twelve knots. This makes the following showing:

STEAMERS BUILT IN THE UNITED STATES FROM 1872 TO 1879.

Class.	No.	Tonnage.
Iron Screw Steamers............................	76	152,088
Wooden Screw Steamers	25	27,563
Total..................................	101	179,651

Referring again, for comparison, to the ocean-going steam tonnage of the world in 1860, we find that it consisted of 338 steamers, with tonnage of 431,000, divided as follows:

Nation.	No.	Tonnage.
Great Britain..........................	156	250,000
United States..........................	52	71,000
All other Nations.	130	150,000
Total..................................	338	431,000

A Good Showing.

Comparing these tables we see that there has been built in this country within these seven years—and years of great business depression—more than twice the tonnage of ocean-going steamers owned by the United States in 1860, and nearly one-half as much as the ocean-going steam tonnage of the world at that time. Our increase in steam tonnage was greater than that of France or Germany, with all their advantages, including that of buying cheap ships in England. Moreover, of the 52 steamers owned by the United States in 1860, nearly all were wooden side-wheelers, not fit for the foreign trade; whereas, the 76 iron steamers mentioned are all screws, of the most improved modern make. Had we possessed these in 1860, we could have thoroughly blockaded our coast, with some ships to spare as cruisers. Of the 179,651 tons, 80,000 are in the foreign trade, the balance in the coasting trade. These 100,000 tons, now competing with the railroads, could in case of a European war at once be put into the foreign trade, leaving the coast business to the railroads and wooden side-wheel steamers.

How We Started.

It should be remembered here that up to 1872, when we began to build this fleet, scarcely an iron screw steamship had been constructed in this country, nor did the rolling mills and ship yards required for that construction exist, at least not in the sense in which they exist to-day. It certainly speaks well for the creative genius of our mechanics and the enterprise of our merchants and capitalists that, under circumstances of so great difficulty—the difficulty of starting a new branch of industry, against that industry already established and maintained by a vast aggregation of capital in a foreign country; of educating labor; of overcoming the timidity of capital, and the effects of a depreciated currency—they should build such a fleet, giving the benefits of employment to thousands who would otherwise have been without it, and putting into home circulation many millions of money greatly needed. What has been the result, and what is the advanced condition of that great industry worth to the country to-day? Through our present facilities we should be able, in case of need, to construct a similar fleet in much less time; whereas, if we had bought those 76 ships from England, and by war or any other cause they should be destroyed, we should have to take our chance to buy over again, and so on indefinitely. Suppose at some such juncture, England (the only nation from whom we could buy ships, if our own yards were closed by adverse legislation) should herself be involved in trouble, and require all her ships, what position would we be placed in? It certainly is prudent to consider this subject in the light of every possible contingency and interest. As it is, through what has already been accomplished the United States is the second iron ship-building country in the world, and the iron ship is built here cheaper to-day than in any other country except England; cheaper, indeed, than it could be built there when her builders had been an equal period in the business, or even ten years ago. Then this point is always to be considered: What guarantee is there that, if all the nations become dependent upon England's builders, they will not raise the price of ships to suit their inclinations and the urgency of the demand? Is it not, indeed, our ability to compete in ship building that makes and keeps the English ship cheap?

THE VALUE OF INDEPENDENCE.

The fact should not be forgotten, that the great nations, with the exception of England, are interested in having more than one country able to supply them with the ships which they require but cannot build for themselves. They know by experience that the position of dependence is always one of uncertainty. How was it with Russia a little time ago, when the war clouds gathered about her? For thirty years she had bought ships from England, thus building up English interests and supporting English workingmen. Her statesmen and people might have said that she was in no danger through her dependence upon a foreign power for ships and tools. But when the difficulty arose, Russia found her enemy in the very ship yards she had fostered. Was England then a reliable, cheap market for her? In her extremity she had to send agents to this country—the only place outside of England where she could get ships for her defence, and after buying she had to go through all legal technicalities and delays before she could get the ships from our ports. It was a costly result of her position. Spain found herself in a similar predicament eight years ago. But there was excuse for the dependence of these countries, since they have not the internal resources for ship building. Would there be any excuse for ours? What if the war had become general, and we had not been able to build ships either for Russia or ourselves?

Suppose England had adopted in 1840 the policy of buying her ships from us, and had continued to buy our wooden ships because they were cheaper than she could build, instead of going to work, as she did, to discover a new material for ships and develop her own resources. When our war came, how could she have procured the ships to take away our trade?

MUCH TO BE GAINED.

During the last twenty years the various foreign governments which do not build ships have annually paid England from $12,000,000 to $20,000,000 for ships, naval and marine. Would it be of no advantage to our industrial interests if we were able to draw one-half that sum each year to this country for American-built ships? Will anybody give a reason why, with a policy to foster our ship building instead of one to discourage it, we should not in a few years be able to compete with England in that business as well as in other lines? If we can build the best engines, and carry off the first prizes from the world's expositions for our agricultural and other machinery, what is to prevent us from becoming a formidable competitor in ship building? Why, the American iron ship is already acknowledged to be without superior. We lack nothing but a well-defined policy that will enable us to enter into the competition. Is it not time that we should have a policy for our foreign trade? Shall we be more blind to our interests than foreigners are? It must indeed seem strange to them that this great country, with its commanding position and resources, should be so indifferent with regard to its rightful place among the nations, and apparently content to let its foreign relations and interests take care of themselves.

Again, on England's control of the carrying trade depends another important fact. Her ships are the road to market for the distribution of her own manufactures as well as ours. Shut off from those markets even temporarily, what would become of her great manufacturing interests? If, then, England were engaged in war, with no safety for her merchant ships under her own flag and no chance to sail them under ours, what an opportunity there would be to put our goods into those foreign markets now held by England, provided we had the ships ready to carry them thither. Is it said that there is no prospect of such an opportunity. What is the condition of affairs in Europe to-day?

An Armed Truce.

The cable brings from London this extract from the leader in the *Times*, commenting on the proposed increase in the German army:

"What is disturbing in the matter is the vivid revelation it affords of the terrible condition of the armed truce in which Europe exists from day to day. By wisdom and firmness, statesmen may avert a collision of these armed forces, but such an achievement will need incessant vigilance and patience. At such a time, England ought to hold herself as free as possible from all unnecessary entanglements in order to be able, if necessary, to make her voice heard at some critical moment when the whole course of European history might be hanging in the balance. Far greater issues to the world are now at stake in Europe than in any other quarter of the globe, and, in deciding them, England may have a still more beneficent part to play than ever she has yet fulfilled. To play it effectively, she must be strong, and she should be at peace."

Surely this justifies some apprehension on our part, since a collision of the armed forces spoken of would inevitably result in advancing the carrying rates, if it did not jeopardize the carrying trade. Suppose England should not be able to hold herself free from entanglements, what a mighty interest we should have at stake then, and to what risks would our millions of dollars worth of ·exports be exposed! The need of being able to go on with our carrying trade, regardless of foreign complications, as we are able to go on with our other business, surely cannot require argument.

Why we Should be Our Own Carriers.

Is it on any conceivable ground a safe policy for us to become dependent upon a foreign nation for our ships? Consider the vast and constantly increasing products we have to place in the markets of the world. We have by many millions of tons more surplus heavy products to be carried long distances, than has any other nation. We exported last year, as stated, over 11,000,000 tons. At the same rate ˙of increase during the next ten years as during the last ten, in 1890 we shall export over 50,000,000 tons. We should require this year, to place ourselves in our true position on the ocean, an outlay of some $75,000,000 to buy ships with ; and each year, with its increased trade, would add to this large sum. In what interest can the man be working who advises us to buy from a foreign builder all these ships which we now need and shall need, if we are to gain the place that belongs to us? How can any American propose, in view of our future, to make us constantly dependent upon outsiders for anything which we have the means and ability to supply ourselves with? Look at the millions appropriated by our government for railroads, canals, rivers, harbors, and other internal improvements. The railroads, moreover, are well paid in addition for carrying the mails. Up to June 30, 1873, government had expended for inland improvements directly intended to build up commerce, $206,897,768.32. All this was well expended; but why did our statesmen stop there? What are the great ocean steamship lines but the continuation of the trunk lines in transporting our products to market? Why should we control those products on the land, and the moment we get them to the seaboard deliver them over to foreigners? By that method we pay the freight for from 1,000 to 2,000 miles to our own people, and for from 3,000 to 4,500 miles to foreigners, when certainly the greater part of it should go to support American enterprise and labor. What would be thought of a proposition to place our trunk lines in the hands of English companies, and have them run under the control of England's government, with her flag hoisted on the cars? Yet we might do that with equally as much reason as surrender to them our products at the seaboard.

An Illustration.

Let me illustrate by hypothesis our position and policy in regard to the carrying trade. Suppose a situation to exist on the ocean similar to that in 1860, but with England as the country about to be plunged into a long and exhaustive war. Suppose her tonnage in the foreign trade to consist of wooden sailing ships ; while the United States, her great rival in commerce and carrying, had already a fine fleet of iron steamships as well as extensive ship yards to construct more ; all built up through liberal aid extended to builders and owners by our government, which thoroughly appreciated the importance of advancing our shipping interests, as a means both to extend our commerce and develop our superior resources. Suppose her merchant marine were driven from the seas by the exigencies of the war and the privateers which we fitted out to help accomplish that very thing, leaving the carrying trade of the North Atlantic open to us for five years. When peace came at the end of that period and England found her trade in our hands, with heavy taxation at home and a depreciated currency ; with a few wooden ships left, in strong contrast to the great iron fleet we had built wherewith to take her trade ; yet knowing, nevertheless, that she had superior resources to ours, such as would enable her to build an iron fleet and regain what she had lost : do you imagine that in such a condition of affairs she would adopt a policy which would keep her mines undeveloped, close what workshops and ship yards she had left, leave her workingmen to shift for themselves, and give her rival peaceful and undisputed possession of what had been obtained in extremity ? A policy that would compel her merchants, if they bought ships, to turn their English bills at forty per cent. discount into gold and send the money abroad to get the ships from us, simply because they could get them a little cheaper ? And this, too, knowing that the effect would be to render her permanently dependent upon us for ships and carrying, to drain herself of millions of dollars yearly, and to make herself poor in order to make her rival rich.

Now, put the United States in the place of England, and the real position in which we stood is not inaccurately described. Had the nation been England, I tell you her statesmen would have done nothing of that kind, nor have had patience with those who urged such a self-destructive policy upon them as the wise and economical course. They would have begun at once an aggressive policy, self-developing and sustaining, to regain what had been taken from them. England's history proves this. Why should our statesmen and people be less alive to the great interests of our country than foreigners are to the interests of theirs ?

THE NECESSARY FIRMNESS.

Having decided upon what was deemed the best policy whereby to build up England's interests, could the cry of subsidy, monopoly, or any other popular catch-word turn her statesmen from that policy? They had the courage to hold their own course regardless of agitation or opposition, and they never hesitated to do anything that promised to promote their power in ship building, but were alert to discover means to perfect and cheapen that work. Thus when it was found that the building of iron ships in Liverpool and London—by reason of the distance of those cities from the coal and iron mines, and the cost of transportation—was unnecessarily expensive, an act of Parliament was passed giving all tonnage dues to a corporation of Glasgow for the purpose of widening and deepening the Clyde, then only an insignificant stream. The width was increased more than double, and its depth from four feet at low water to twenty-four. Yet these tonnage dues belonged to the government as much as the gold locked in its exchequer. They were given without a word to cheapen the cost of iron ship building, and thus encourage still further the growth of that fleet to which England's representatives looked for her supremacy on the seas. Here was an example of subsidy and monopoly for you to add to that of building naval ships in the private yards! But suppose that howl had been raised, and had succeeded in frightening Parliament from its position. That would forever have kept as a mere mudhole that river now known the world over, where, I may almost say, the navies and merchant marine of the world are built. Has England anything to regret for the adoption of such a policy of liberal encouragement? Her commercial and carrying supremacy and the millions on millions of gold annually drawn to her treasury make the best answer.

THE VAST DIFFERENCE.

But how is it when the American capitalist invests his millions in ship building on the Delaware, convenient to the mines, wanting neither drainage nor dredging, running from Trenton to the breakwater, a distance of 130 miles, open to the whole country and its capital (even to foreign capital), every part of it well adapted for ship building, with room enough for all the ship builders of the world? Why, he is at once assailed by the advocate of free ships; the cry of Monopoly is rung in the ears of the nation; that "monstrous indefinable shape" called Subsidy is conjured up, and some of our public men are deterred by such means from taking the action they would like to take, from advocating measures which foreign statesmen never feared to enforce.

The American builders on the Delaware may have converted the swamp into a hive of industry; established ship yards of national importance; given employment to many thousands of American workingmen who have thus been enabled to provide themselves with comfortable homes, churches, and schools; and paid both national, State, and county taxes. Yet the advocate of free ships goes on year after year threatening the very existence of their capital, and trying to bring ruin upon the men who had the courage and enterprise to use it in spite of all obstacles. If the ship builders and owners are to be crushed out, let them know it. If not, let the free ship men know that, and let intelligent attention be turned to the modification of existing oppressive laws, so as to give the American builder and owner some chance to compete with foreign rivals. That would certainly be no more than justice.

XIII.

FREE TRADE.

The theory of free trade is doubtless approved by all men. It would be a grand result to bring about such a condition of things universal that we might see the nations of the world enjoying free commercial intercourse with each other. America now practices free trade to a greater extent than any other country, after all, for here we have this great community of United States—covering an immense territory, with a population of 46,000,000 of people governed by the same laws, standing before the same tribunal—all having free trade throughout their length and breadth. When the many nations become like our States, under one law and government, and pursuing a similar policy for the elevation of man, then there may be free trade with all the world. But until that millennial period arrives free trade must be looked at practically as well as theoretically ; and the true friend of free trade is the man who first studies thoroughly the existing condition of affairs, and begins by advocating a policy which will pave the way for free trade.

The practical change can never be brought about suddenly. All things must be made equal in the competition, before the doors are thrown open to the world. The material, the capital, the manufacture, and the labor must be as cheap here as they are elsewhere before we can maintain our industries under free trade. If the free trader is willing that our people should give up these industries, and that we should return to and remain a nation of cotton and bread growers merely, then there is nothing more to be done with him, and he is advocating the policy that will accomplish that very thing, provided our people can ever be brought to adopt it.

The late Napoleon made the blunder of his life through not understanding the condition of the military power of his own nation or that of his rival Germany, over-estimating the one, under-estimating the other, and beginning war before he was prepared. He met in consequence with ignominious defeat, and humiliated his country and himself. The believer in free trade certainly does not wish to make a like mistake. He should carefully consider the resources, development, policy and trade of the countries with which his own has to compete, recognize their rates of labor and capital, and make sure that his country is prepared to meet her rivals in the competition. To make the experiment and fail would bring ruin upon the land. Theoretical experiment in such cases is too costly. We want to know the consequences before we act. Free trade in ships would include free trade in the many branches of industry connected with the building of ships. The same principles of trade apply to all our industries, and the observations upon this subject are therefore made general.

NECESSITY OF EQUAL DEVELOPMENT.

It is evident that the solid prosperity of a nation depends upon the development of all its resources, as equally as possible. The nation that can within itself supply most fully its own wants is the most independent and prosperous. England to-day, with all her power and wealth, and the great development of her manufacturing interest and the ocean carrying trade, feels her weakness in the inability to produce bread and cotton. To make up for that inability, her policy has been and is to do what she can to confine us to the production of these raw materials which she requires, but cannot by any policy or legislation produce from her soil; then to carry these products in her own ships, and pay us for them in the manufactured articles, also sent over to us in her ships. This is certainly a sharp and profitable policy for England, but would it be well for us to fall in with it? The best market we have ever found is our own, and we required no other until recently. Now that we over-produce, will it aid us in disposing of our surplus to destroy the home market?

Yet that is what the free trader proposes to do, through his onslaught on our manufactures. To illustrate, divide our entire population into two classes: 1. The agricultural; and 2. The manufacturing, mercantile, and professional, including all the varied occupations outside of farming. Now, reduce the manufacturing population to the condition of the European mechanic and workingman, with the wages they receive, and you at once lessen their power to purchase. This, of course, affects the grain, cotton, and meat producer, since evidently you cannot get as much for sixty cents (which is a fair proportionate estimate of the difference between labor here and in Europe) as for a dollar, unless you reduce the price of the articles; and reducing that price must lessen the producer's income. This great home market is of the utmost importance to us, and it has been well said that "the more we stimulate and increase it, the better it is for the agricultural as well as every other interest in the country. Protection does this; it sustains the manufactories, thereby making a market for the farmers." The free trader, however, puts before the farmers of the West and South another phase of the question, which better answers his purpose. He cries monopoly, and advocates a sectional policy. He calls upon the farmer to think of the advanced cost at which he must buy his plow-share, his ox-chain, his wagon tire and horse-shoeing; but never shows him that, as a consequence of paying a small advance in price for what he purchases, he has a home market where he gets in return a much better price for his products than he could if the mechanic and artisan were paid less wages. Nor does the free-trader ever suggest to him that the advance in cost above other markets is in the interest of American labor. On the contrary he tells him the absurdity that it is in the interest solely of American capitalists. Let the people see for themselves. They will find that the result of the increased pay to American labor is, that there are no other people in the world so well fed, housed, clothed and educated as are our own people of all occupations.

The Value of Cheap Money.

The high rate of interest in this country, where we have not yet accumulated vast capital, is a most important point to the American manufacturer, whatever he makes or builds, be it cotton thread or iron ships. Suppose two men are engaged in manufacturing the same article for the same market. One of them pays seven per cent. for his capital, the other three. Would the competition be fair under free trade? and how long in these circumstances could the manufacturer who had to pay seven per cent. continue to manufacture? The free trader should first secure the same low rate of interest to both manufacturers, and thus give them equal chance to compete.

Let some advocate of free trade take a paper, call upon our capitalists, and ask them to subscribe the capital to start a cotton factory, rolling mill, blast furnace, or ship yard. Before he can get a dollar subscribed the questions will be asked: "Will it pay?" "Who have you to compete with?" When he answers that his competition is with men who get their capital for three per cent.,—that is, four per cent. less than he can get it—and their labor for forty per cent. less; who pay less taxes, have good business reputation, and are well established—what chance has he, think you, to obtain the capital desired? Capitalists do not invest in enterprises under such conditions. But let there be such legislation as would enable the capitalist to invest profitably, and induce him to do so, and this same free trader, instead of taking the opportunity offered him to get capital and start in the business of developing our industries, would very probably stand one side and cry out, "Monopoly, monopoly."

A POINT WORTH NOTING.

But for our wise legislation during the war in protecting the manufacturing interests of the country at a time when we were struggling under the war's burdens, foreign capitalists and manufacturers would have closed every one of our factories and workshops, and utterly drained the country of its wealth. Thus a second of the three great sources of prosperity—manufactures—would have gone from us in the same way our carrying trade did, and at the close of the war it would have been as hard to find a factory on the land as it was to find an American ship on the sea. And in that case, when our merchants and manufacturers showed disposition to rebuild and make a new start, they would have met with the same spirit of opposition from a kindred source to that whence now comes the advocacy of free ships. If you watch to-day the tone of English sentiment you will find it in harmony with the quotation previously made from Mr. Wilson—the view being that it is better for us to give our whole attention to growing cotton and corn, and barter these products with England for all our other wants.

LET THERE BE NO DISCRIMINATION.

On what ground of right could free trade be applied to ships, while protection is given to other industries requiring it not one whit more? The principles of the free ship bill now before Congress, if applied to all other of our industries, would bankrupt nine-tenths of the business men of the country. Assuredly the interests of all citizens are alike sacred under our laws. What right would there be in taxing the American ship builder's yard, tools, and machinery, his invested capital, and even his workingmen, to support our government, while allowing at the same time a foreign ship builder, paying taxes to support a foreign government, to compete with him and his home-built ships; it being known, moreover, that the foreign builder had advantages of cheap interest and labor which the American could not obtain? If to get the ship cheap is the whole object, as the free ship men profess, would it not be a far better policy to remove all taxation from the American built ship, and thus at least keep our money at home for the benefit of our own people? If the ship is built abroad under the free ship bill, this country would get no revenue from that source.

BUYING AND SELLING.

On the point of barter there is, it appears to me, a vast deal of nonsense uttered. For years the balance of trade was against us. Last year the crops were poor in nearly all Europe, while in this country they were better than the average. Therefore, in bread and meats we sold abroad large quantities of our products at good prices. At the same time the improvements in the machinery used in our manufacturing industries enabled our manufacturers to enter the market with such goods and prices that our people bought more largely than usual of our own manufactured goods. In this way we both kept our money at home, and brought into the country large sums of gold from Europe, turning the balance of trade in our favor, and giving our currency a solid foundation. No sooner did this fortunate turn take place, however, than much anxiety was manifested on the part of some of our papers and people as to how our foreign customers were going to find money to pay for what they bought from us, if we did not buy more from them in return. The anxiety could scarcely have been greater had these papers and people belonged in London instead of in the United States.

Now I cannot learn that England, in all the years in which the balance of trade was against us, troubled herself in the slightest as to how we should find the money to pay for what we wanted to buy, though our dollar was worth only seventy-five cents or less, and had to be converted into gold for her benefit. Nor can I learn that these people and papers in that critical period ever made any outcry against England for not taking more of our products in exchange for our purchases. The fact is, no matter what economical theorists may proclaim, that this matter of barter is neither one of friendship nor favor. That we must buy in order to sell is the purest nonsense in the form in which it is put forth by our free trade clubs. If we have goods to sell which another nation requires, she will buy them from us if she can get them here cheaper and better than elsewhere. If she can get them to better advantage otherwheres, she will do that. And we need not trouble ourselves as to where she will get the money, so long as the European nations have plenty of it at three per cent., while in our Western and Southern States, where the bulk of what we sell abroad is produced, money costs the producers from seven to twelve per cent. It is a point to be considered, too, that we sell England bread, meat, cotton, and petroleum—necessaries. A country which must have these staples will always find the money to pay for them, when she cannot find money to pay for diamonds, laces, kid gloves, and other luxuries.

A Matter of Price.

On this question of exchange an American writer in a recent letter to an English free trader aptly says :

" There is an error which most of you Englishmen fall into when discussing this matter, viz : That what you buy from us, depends on what we purchase or take of you. In other words, if we do not purchase your manufactured goods you will not buy agricultural products from us. Now there was never a greater fallacy. If you can buy your grain and breadstuffs in Russia cheaper than you can in America, you buy them there ; if on the other hand, we can sell to you at a cheaper rate than Russia, you buy of us. It is price that regulates and controls, and not the balance of trade between the two countries. Do you suppose that any grain dealer in England ever looks to see whether the balance of trade is for or against his country when he is about to make a purchase? He buys wherever he can obtain the grain for the lowest price. As proof of this take the trade of your own country with Russia for the last twenty years. There has been not one single year during this period in which you have not purchased of her greatly in excess of (and in most years more than double in value) what she has bought of you. Your trade with Russia for the last twenty years was in the aggregate as follows: Your imports were £369,782,059, and your exports £158,436,122. In other words you buy of Russia more than double what she buys of you. And if you will examine the statistics of your trade with other foreign countries you will find the same results—proving that what you buy of a nation is not dependent upon what she buys of you ; that it is price and not the balance of trade that regulates and controls the business you do."

Surely we need not be afraid to get a little surplus gold into the country. The crops in Europe next year may be good, and in that case a smaller quantity of our breadstuffs will be required ; while even if the same quantity were taken, the price would be reduced. Considering how short a time we have had the balance of trade in our favor, we have no cause yet to waste any sympathy on our customers in Europe. When we have had that balance for some years, and can furnish capital at as low a rate as France and England can, it will then be time to think about suffering foreign interests. Is not much of our present prosperity due to the bringing of this gold here and keeping it here ? What would have been the effect upon our financial condition if we bought as much of England's manufactures last year as in former years?

It may be pertinent to inquire, here, how long England could continue to furnish capital at three per cent. if she were compelled to drain herself of millions of gold annually to pay for ships and freights, instead of being able from these two sources to draw gold to her treasury from all other nations. And if we were getting our fair share of the carrying profits, in addition to the changes in trade, would not money with us reach an equally low and favorable rate of interest?

A Point and a Question.

The most radical free trade advocates among the English writers have always admitted that in the early development of an industry, where the natural resources existed in abundance, a certain protection was wise and essential, even at some sacrifice to the people at large. England's delay in the repeal of her Corn Laws, for example, has been justified by her free traders. But our free ship men are not even willing to allow that much. They started by introducing a bill to prevent capital from either coming here or being collected at home to be put into ship building, before our iron ship building had opportunity to make any headway, and they have fought against it ever since.

How will the free traders explain the fact that Germany, after giving free trade a certainly fair trial, with cheaper capital and labor to help her in free trade than we have, has been forced to return to protection; and why is England so bitterly opposed to this return by Germany to a protective tariff? I know of no argument that should cause us to think more deeply before we act in so serious a matter, than the unquestionable fact that England, depressed herself under free trade, is so eager to have us adopt that policy.

XIV.

THE LABOR PROBLEM.

Foremost in importance, as the vital point of this whole subject of American industries, is the labor problem. Whatever the branch under consideration, the fact must be met that a large percentage—from sixty to ninety—of the cost of manufactured goods is labor; that this labor is from thirty to fifty per cent. cheaper in Europe than in this country: hence that the cost of production there is less than it is here. To render competition fair or possible under free trade, therefore, the price of labor must either be advanced in Europe or reduced in the United States. This is a part of the subject which the advocate of free trade carefully avoids. He sees in the protective tariff nothing but profits to the capitalist and loss to the masses. It will be worth while to look into this somewhat, and discover whether that view is the correct one, or whether the truth is that protection is in the direct interest of labor and of the whole people. It is imperatively necessary in a country like ours that this matter should be rightly understood, and the character and benefits of the American labor policy be appreciated by all.

LABOR IN THE SHIP YARD.

Let us consider first the relations of labor to ship building. Having more coal and iron (and iron of better quality) than any other country, and having these essential products easy of access and close to the great streams where ships can be built; having the rolling mills to make the iron in the very yards where the ships are built; having the most energetic, intelligent, and capable class of workmen in the world—will the advocate of free ships tell us why we cannot build ships here as cheap as they can be built in England? In answering that question correctly the facts which follow will be of assistance.

Take a great iron ship yard, with capacity to employ 2,000 men. To conduct this, with the grounds, buildings and tools required, involves an outlay of $1,500,000 capital. In England the interest on that capital is $45,000 ; in this country it is $105,000. Here is a difference against us of $60,000, which certainly cannot be counted as profits of protection for the capitalist. What are the wages paid these 2,000 men in English and American yards respectively? The following table, taken from Young's "Labor in Europe and America," will show. It gives the average cost of this class of labor in the two countries in 1874, the latest date to which the comparison has been brought. The figures represent the weekly wages of one workman in each department of the ship yard :

COMPARISON OF THE RATE OF WAGES PAID IN THE UNITED STATES, ENGLAND AND SCOTLAND TO THE DIFFERENT CLASSES OF MECHANICS ENGAGED IN THE CONSTRUCTION OF AN IRON SHIP AND HER MACHINERY.

	United States.	England.
SHIP YARD DEPARTMENT.		
Ship Smiths....................................	$15 95	$6 05
Angle Iron Smiths.............................	13 20	6 29
Helpers..	8 80	3 75
Riveters.......................................	11 00	5 20
Platers and Fitters...........................	13 20	6 40
Calkers..	9 35	5 32
Laborers.......................................	7 70	3 38
Rivet Boys	3 30	1 69
Carpenters and Boat Builders..................	13 20	6 53
Joiners..	12 65	6 53
Painters.......................................	12 10	7 32
Riggers..	11 00	6 20
Planers..	8 80	5 68
Punchers.......................................	8 80	5 00
STEAM ENGINE DEPARTMENT.		
Draftsmen......................................	19 80	8 22
Pattern Makers.................................	14 30	6 41
Engine Blacksmiths.............................	13 20	6 59
Helpers..	8 80	3 87
Finishers......................................	13 20	5 86
Turners..	13 20	6 05
Planers..	13 20	6 25
BOILER DEPARTMENT.		
Fitters..	13 20	6 47
Riveters	11 00	5 44
Calkers	9 35	5 00
Holders on.....................................	8 25	4 00
Laborers.......................................	7 26	3 86
Boys, Heaters and Passers......................	3 30	1 25
Flange Turners.................................	16 50	6 20
IRON FOUNDRY.		
Loam Moulders..................................	16 50	6 50
Green Sand Moulders............................	13 20	6 87
Melters	13 20	6 00
Helpers..	7 70	4 00
BRASS FOUNDRY.		
Brass Moulders.................................	14 30	6 15
Melters..	8 80	5 50
Chippers.......................................	11 00	4 00
Laborers.......................................	7 70	3 75
Total weekly wages, 36 men....................	$406 01	$192 60
Weekly wages of 2,000 men, at same average.......	$22,540 00	$10,700 00

From this table it will be seen that there are in the ship yard, using the raw material after it is made into shapes, thirty-six different classes of mechanics. Compare what the American workman of each class receives per week for himself and family, with the sum received by the English workman of the same class. If you could look into the homes of these men on pay-day the difference would be still more apparent. As it is, a study of the column of wages paid to skilled labor in England will give an interesting insight into the condition of the workingmen there.

What Makes the Cost.

These figures ought to show with unmistakable plainness where the difference in the cost of the ship lies. Ninety per cent. of that cost is labor. The men employed in that labor in this country were paid in 1874 more than twice as much as their fellow workmen in England. The difference to-day in the cost of that labor, to make a low estimate, is at least forty per cent. in favor of our workingmen. Through the advantages of our position in respect to raw material, and the superiority of our mechanics and methods, we have been able to reduce the difference in the cost of the ship from forty to twenty per cent. With these facts before us, realizing fully that ninety per cent. of the total cost of this product—from the ore in the mine and the timber in the forest to the completed ship—is paid to labor, in whose interest is the protection that makes it possible for us to build this ship? Is not the conclusion irresistible that it is in the interest of the workingmen to whom it furnishes employment? What truth or justice is there, then, in the cry of monopoly uttered against the men who keep this labor employed? The specious assertion that protection merely puts large profits into their pockets is sufficiently answered by the figures above, which show that where the English builder pays $10,700, the American builder must pay for similar work $22,540, and every cent of it to the men whom he employs. Surely that is not the way in which large profits are to be gained by the capitalist.

Two Methods of Solution.

It may be regarded as proven that if we are to take on the ocean the place which belongs to us, we must build our ships at home, not buy them abroad. The question, then, as to whether we shall strive to be first among the carrying nations, or be content with third, fourth, or fifth place, resolves itself into the labor problem. This may be disposed of in one of two ways:

I. By reducing our working people to the condition of those in Europe. Under the depression caused by over-production of manufactured goods consequent upon loss of trade, and various other adverse circumstances, England has ground down her laboring classes until their condition is more pitiable now than it has been at any period within forty years, and such as I hope our working people may never reach. In all branches of industry the situation has been equally hard for labor, and the discontent has more than once assumed such proportions as to cause serious apprehension among her statesmen. Nor has this dissatisfaction of labor been confined by any means to England. It may be said to be general throughout Europe, and the results are either outbreaks or wholesale emigration. Now there is no doubt that if we can reduce our labor to a like condition, we can not only build ships as cheap as the English builder, but also manufacture all other articles as cheap as any other nation. Such reduction of labor would settle all the questions of the tariff, but it would at the same time raise up others far more troublesome. It is not in accordance with our national policy, moreover, to compete with our rivals by following their example of grinding down the working classes. Our system of government was framed in the interest and for the elevation of man, and we have never yet abandoned the original idea of its founders. We are to bring other nations up to our standard in this respect, not lower ourselves to theirs. Until they adopt our principles, and place labor upon a higher plane by properly compensating their workingmen and increasing their advantages, competition with them is impossible except under the protection of our just laws. Surely no American would advocate a policy of retrogression on our part.

What does it mean to lower the price of labor? Does it not mean to take from the American workingman's table the better food, from his back the better clothes, from his family the more comfortable home, from his children the superior advantages which are now enjoyed by him and them, as compared with the condition of the foreign workingman and his family? Does it not mean extreme poverty, ignorance, dissatisfaction and emigration? Introduce this policy here, and not only would the hope of the American workingman be crushed out, but the hope also of the laboring classes throughout the world. For the example and influence of this free land have given a new ambition and impulse to man in every part of the globe. It was in good faith, believing in this spirit of our institutions and in the maintenance of our people in their improved conditions, that our capitalists invested their money in enterprises for the development of the country and the employment of its people. They certainly did not expect to encounter legislation that would throw them into competition with the laboring classes of governments whose policy is totally unlike our own. Would the free trader like to assume the responsibility of the attempt to thrust the workingmen of this country back into a condition out of which they have either grown, under favoring circumstances, or emigrated? Would any political party like to lead in such a movement? The European nations require their great standing armies quite as much to protect them from internal uprisings on the part of oppressed labor, as from outside enemies. In the hard times following our war, despite the improved fortunes of our working people as compared with those of foreign countries, we had sufficient experience to show that a vast army would be required to preserve peace within our wide territory should the grinding down method be adopted. Nor would the great masses of the people permit any such policy. As a mere economical question, moreover, which would be wiser, to protect our industries and pay the slight difference in cost of home manufactured products, so long as may be necessary ; or to pay the cost of a standing army with which to protect capital against labor driven to the wall ? It is plain that this way will not do. The second way is :

II. By wise legislation, and a policy broad and liberal, befitting this powerful nation. Let our statesmen do one-half what England's statesmen have done to build up her ocean carrying trade, and we are surely destined to become the first carrying nation of the world, despite the thirty years start England had of us in iron ship building. Our history from 1789 to 1830 shows what we did in building wooden ships, the ocean carriers of that day. We can make a like advance and obtain a like position through iron ship building if we are given a fair chance. Why should we not have it?

<h2 style="text-align:center">The Iron Business.</h2>

The same principles which obtain in regard to ship building apply to all other of our industries. It is a question of labor and home development throughout. Look at the iron business, just now prominent by reason of the advance in price, and closely connected with ship building. Our resources in coal and iron are superior to those of any other country, yet iron is dearer here than in other countries where it is made. How is this to be accounted for? Suppose that to make a ton of iron it takes say two tons of ore and one and a half tons of coal. The coal and ore in the mine are considered of equal value in both countries, and the three and a half tons may be valued at $3.50. The large difference between that bottom cost of the raw material and the cost of the finished iron in this country and in England is due to the difference in the cost of the labor—forty per cent. less in England—which works up that ore into iron. In this case, as in that of the ship, the protection which keeps the iron industry alive in this country is in the interest of the workingmen to whom it secures their superior wages. And the iron, from the pig to ship-plate or watch spring, is but increased labor, requiring the same protection in behalf of the American workingmen.

It is a fact worthy of attention that the demand for iron in this country above what we produce regulates the English iron market. What is the cause of the recent rise in the price of iron? The low prices for some years forced many of our furnaces out of blast, as it was unprofitable to run them. The resumption of specie payments and our reviving prosperity created a sudden demand for iron. Our own works being unable to supply immediate wants, the demand falls upon the English market. What is the result? As England is the only exporter of iron, the moment we need that product she raises its price ; and there is no way to prevent this, since we must have the iron and cannot make it fast enough ourselves. Thus, within three months the price of iron in the English market was raised forty per cent., and under our demand the English iron industry, long depressed, has been revived. The question is important—if we had to rely still further upon England for our iron, to what extent would the price be advanced ? Before we get cheap iron again, the furnaces which were closed here during the depression in business must be put in blast and the demand be supplied at home. England certainly will not help to cheapen the price.

Iron and Civilization.

It may well be asked, in this connection, whether we can afford to give up the development of our own resources, either in iron or in any other product. What would have become of our Confederacy of States but for the iron bands that have bound these States together. With the slow methods of communication which we had before the era of the railroad, steamboat, and telegraph, this Union could never have been maintained as the States grew in population and power. Unable to come together without great expense and loss of time, whether for legislation or trade, the people of the various sections virtually shut off from each other, their interests would have grown apart, and we should have had a system of petty governments. It is safe to say that iron has done as much for the advancement of the United States in greatness and civilization as legislation and all other forces combined. To depend upon a foreign country for a material which has done so much for us as a nation would be worse than folly. Count the enormous drain it would have been upon our resources had we sent abroad the cash to pay for all the iron we have consumed, instead of developing our own mines and the iron industry. Look at what iron and coal have done for England. With a territory not larger than two of our States, without agricultural resources to supply food for her people, how has she gained her proud eminence among the nations? Through her iron and coal, and the devotion of her talents to their development, she has made herself first on the ocean, the chief manufacturing nation of the world, and able, if not to dictate the policy of Europe, at least to wield powerful influence in shaping that policy.

We have every advantage possessed by England, and even greater ones. We have the coal and iron, the skilled labor, and the brains to employ all our resources. Shall our statesmen lead us to the first place, or shall we remain where we are? How can any American propose to put a check upon the development of such a mighty industry as this at home, and at the same time give new impetus to that industry in a foreign land? The same question will apply to any of our industries. Which shall we first build up, our own or a foreign land? But no sooner are efforts made to advance our own interests than the old cry is heard—Monopoly.

A Queer Kind of Monopoly.

Where is this monopoly, when you come to search for it? Here are 4,000,000 square miles of territory, 46,000,000 of people, inexhaustible resources. Are not the mines, the river banks, all our resources and facilities open to the capital, not alone of our own people, but of the world? What law is there to prevent anybody from investing in any industry or manufacture whatsoever in this country? Everything here is free, both to the capital and labor of all nations. Where in the name of common sense is the monopoly? If there be any, it is, as I have shown, accorded to our working people. During the depression of the last seven years no class of men lost and suffered more by failures than did the manufacturers of our country. In proportion to these the middlemen and free traders lost little. Yet the closing of a single one of our great workshops which was engaged in developing the resources of our country, was more of a loss to the country than the failure of a dozen importers of kid gloves and fine laces.

This outcry about monopoly is one of the most shallow, untrue, and unjust means employed to work injury to our manufacturing interests. Why is it that foreign capitalists do not come here to invest in iron mines or ship yards? Not because they fear any monopoly, but because they have seen the agitation of this tariff question every year in our Congress, and know that if they were to put capital here, a free trade wave or any large reduction of our tariff might at any time sweep it away. Trained and shrewd capitalists have no desire to invest their money where it will fall between the millstones of the free traders and the labor question. If there is any such thing as monopoly in this country to-day, it is due more than any thing else to the annual exhibition of tinkering with our tariff, which effectually prevents the investment of fresh capital here. As business men, you know that capital will never be invested in a thing to which one value may be given this year and another next year, by legislation. I believe that, if the capitalists of this country and of Europe were certain that for ten years there would be no tariff legislation outside of changes for revenue, millions of money would come here from abroad for investment, and that our own capitalists would also invest their millions in our industries, the result being a sharper competition, increased development of our resources, and cheaper production. Beyond question there are millions of American capital, now invested in bonds and other securities, which would be quickly put into manufactures if the tariff question were settled.

The Proper Thing to Do.

If the free ship man or free trader sincerely wishes to cheapen the price of the ship and other products of our industry, why does he not begin at the right place—at the ninety per cent. instead of ten? The ten per cent. in raw material is as cheap here as in any other country and needs no attention. It is the ninety per cent. that gives the trouble. But the free ship man will not touch the real difficulty. In place of that he shapes his argument to divide the people by sectional appeals, and create jealousies between capital and labor. If these men were honest and understood the subject, if they wanted to attack the source of the difficulty, every free trade club would at once send out missionaries to convince the working people that they are enjoying more of the good things of life than they are entitled to, and that they ought to be willing to take less wages in order to enable us to compete with England in free trade. Any party in politics or free trade that will take this stand honestly and boldly, thus proclaiming their true policy, will soon be convinced that such a doctrine would receive little consideration from the masses of the people. This country is not yet a community of aristocrats. The majority of our leading men throughout the country came from the workshop and the plow, and are not likely to forget that they owe much of their success to the principle of our land which recognizes the dignity of labor and seeks its elevation.

XV.

THE NEED OF NEW MARKETS.

The opening up of new markets is a subject closely connected with our position as ocean carriers, and involved in our shipping policy to an extent not yet appreciated by our people. The capacity of this country to increase its production beyond the home consumption, and even beyond what Europe requires of us in ordinary circumstances, is a matter of vital concern. It is not only worthy, but demands the careful consideration of our public men, and it cannot long be put aside. Look at our normal increase in surplus products :

From 1869 to 1879 our exports grew from 2,482,172 tons to 11,149,160, or more than 400 per cent. The following table shows the rates of increase in the various departments in that period, and the estimated increase in the same proportion during the next ten years :

Products.	1869.	1879.	1889.
Agricultural	1,404,642	7,947,230	44,900,000
Provisions	104,784	699,430	4,664,500
Manufactures	198,605	190,600	600,000*
Oils	455,857	1,564,600	5,360,000
Metals	318,283	747,300	1,563,852
Total	2,482,171	11,149,160	57,089,352

*The probable increase in our manufactures cannot well be estimated, as it will depend almost entirely upon our success in opening the markets which demand those products, and hence upon our policy in regard to the carrying trade.

This shows that while we have advanced rapidly in every other class of exports, in that of manufactures there was a slight falling off. One reason for this, and an important reason, was that we had no means of direct and rapid communication with existing markets that would have taken our agricultural products, and together with these large quantities of our manufactures. Foreign nations had such, communication, while for want of it we lost markets naturally ours. That is chief of the things we have to remedy.

The great increase in the exports of agricultural products and provisions tells a story of deep interest to the people of the West and South—the great agricultural and grazing sections. When they can get money at three per cent. they can undersell all competitors in the European market. Hence no part of the country is more concerned in every means which tends to bring all the money possible into the country, and so cheapen it; and no part should feel more lively concern in the question of the revival of our carrying trade, and in what follows as a logical consequence—the opening up of new markets for our increasing products of all classes. This is beginning to be recognized as the great question of the present and immediate future.

In comparison with the actual resources of our country in grain, for instance, what we now raise, gather, and send to market is but the chicken feed. Suppose, as we have in the above estimate, our increase in the next ten years to be equal to what it was in the last decade,—and it will doubtless be even greater, since we have had to contend against a depreciated currency and conditions of great depression, and did not produce nearly as much as we had capacity to produce—what shall we do with this vast surplus? It has been said truly, that just so long as we can dispose of this surplus at reasonable prices, just so long will our prosperity continue. The welfare of the country, nay, its salvation from such conditions of depression and dissatisfaction as now rest upon Europe, depends upon the timely opening of new markets.

OUR OPPORTUNITY.

The great South American markets are open to us naturally and legitimately, and offer us a vast and profitable field for competition in trade. Upon this point the New York *World* of May, 1877, well says :

" The United States are fitted to occupy the leading position in the trade with South America, both by nature and the energy and inventive genius of the people. South America produces only a fraction of the amount of the necessaries of life which her people consume. She finds profitable employment for her people in raising purely tropical products, cotton, coffee, india-rubber, etc. South America accordingly does now and will for years go abroad to buy the greater part of her food, clothing, furniture, building materials, etc., which she would rather buy than produce. All these things, or nearly all, can now be bought in the United States as cheaply as anywhere in the world ; but American merchants have simply neglected the market, and the consequence is that nine-tenths of what the South Americans import is shipped to them by Europe from points 1,000 to 5,000 miles further away from them than the ports of the United States."

The following table shows the unequal character of our export and import trade with the principal South American countries, and the natural chance we have there to open up new outlets for our surplus products. The figures are for 1875, since which date the proportion has not materially changed :

TRADE OF THE UNITED STATES WITH SOUTH AMERICA.

Countries.	Imports.	Exports.
Brazil	$42,033,046	$7,634,865
United States of Colombia	12,942,305	4,272,950
Argentine Republic	5,834,709	1,301,294
Uruguay	2,935,039	1,440,665
Venezuela	5,690,224	2,423,254
Total	$69,435,323	$17,073,028

Let us look a little closer into the trade of these South American nations, in which we are so deeply interested. Of Brazil's total exports of $101,350,000 in 1870, we bought nearly one-half. Yet of her total imports of $83,350,000 we sold her only about one-eleventh. In contrast to this, England bought from Brazil $30,000,000 worth of her products, and sold her more than $28,000,000 worth of English goods in return, thus buying one-fourth less than the United States, while selling four times as much. France and Germany, in taking the remainder of the exports, were favored in the same way to still greater extent. Take the Argentine Republic : while we buy from her five times as much as we sell to her, England has a balance of trade in her favor as against that country of over $8,000,000, and France a like balance of $2,000,000. Uruguay has an export trade of some $15,000,000, of which England, France, and the United States take about one-third each. She has an import trade of $16,000,000. Instead of contributing one-third in return, we sell her our products to the value of $1,440,665, and allow England, France, and Germany to sell her the $14,500,000 odd balance. To give one more instance, the United States of Colombia sells us her products to the amount of $12,942,305 ; while we sell her but $4,272,950, or one-fourth as much of ours in return. These facts ought to show plainly enough what trade fields there are in South America for us to cultivate. On the free traders favorite theory of the reciprocity of trade, there is every ground of reason why we should enter in and occupy these markets in lively competition with the present foreign occupants.

Our unequal trade with the South American countries has gone on for years ; yet I have never known the free traders on that account either to express anxiety as to where we should get the gold to pay those countries for what we buy from them, or to explain this flat contradiction of their doctrine that a nation will buy only where it sells in exchange.

THE GREAT MARKET OF BRAZIL.

The chief product which we buy from Brazil, coffee, is one that we must have, but cannot raise in any part of our own territory. It is not necessary for Brazil to send it to us. We must go and get it, and pay our gold for it, unless we can induce her to accept something else in exchange ; and hitherto, even in the getting of it here we have unfortunately had to call upon English ships to carry it for us. As we increase in population our demand for coffee increases proportionately. Brazil wants, in turn, our bread, lard, ham, and other food, as well as clothing and all lines of manufactures. She will never become a manufacturing country, and will purchase manufactures as well as food from other nations. No nation is able to supply her with everything she needs to greater advantage than we are. With her 13,000,000 of people who, I might say, have to be fed and clothed in exchange for her natural products, there is no new market more important to be secured than hers. We are her chief customer, taking straightway one half of her total surplus products. She must buy somewhere the products which we have to sell, and have in such increasing abundance that we must either dispose of them or suffer ruinous depression in every department of growth and trade. Shall we supply these people with what they need?

It should be remembered, also, that Brazil has 3,200,000 square miles of territory in which railroads are to be built as the country is developed. There are but two nations which are able to construct those railroads—the United States and England. You may be sure that we shall not build one of them if English capital and influence can stop it.

Here is an illustration of the interest Great Britain takes in preventing American enterprise from getting a foothold in South America. Some years ago an English company undertook to build a railroad in one of the South American countries. The difficulties of the construction in a wild, unsettled, and unhealthy region of territory were so great that the job was abandoned in utter discouragement. An American company then proposed to go on with it. The capital wherewith to build the road was on deposit in London. The Americans took hold with a will, and accomplished wonders for the length of time they were able to continue. They were allowed to proceed until they had over half a million of their own capital invested, and were certain of success so far as the construction was concerned. Then by a series of litigation their right to spend the funds deposited for building the road was contested in the English courts. At first the suits were decided in their favor, but the case was carried up from court to court until, after tying up all their capital, the American contractors were compelled to bring their men home and give up the work. Nor have they got their money back yet. This was the kind of warning served by the English capitalists to American enterprise in South America.

Why is it that, with so many circumstances in our favor, our trade with South America is so one-sided? Simply because until recently there has been no direct and rapid communication between the merchants of South America and the United States. This matter is certainly worth looking at in some detail, and it furnishes a most interesting if not over-complimentary view of our foreign policy as compared with that of other nations. If we have been as neglectful of our commerce as of our carrying trade, it must not be expected that the foreign governments have either followed our example, or failed to profit by our course.

Brazil's Mail Contracts.

The official record of Brazil shows that for the year ending June 30, 1880, the Brazilian Congress appropriated $1,507,000 for mail pay to steamship lines. This sum was divided as follows:

Mail pay to English lines........................	$1,097,000
Mail pay to Brazilian lines.......................	310,000
Mail pay to the New American line...............	100,000
Total.....................................	$1,507,000

The English lines certainly have the lion's share in this case. They comprise not only the lines from Rio to Liverpool and Southampton, but also lines engaged in the coasting and river trade, and connecting with the ocean lines. Thus the English own the Bahiana, San Francisco and Bahia, Rio and Para (paid $404,000), Rio, Montevideo and Buenos Ayres (paid $120,000), coast of Brazil and River Plate, Cayaba and Montevideo (paid $150,000), Rio and Rio Grand de Sul, Rio and Santa Catharina, Liverpool and Amazon River Line (paid $240,000), and other lines. The $100,000 now paid to the American line for carrying the Brazilian mails to New York were formerly given to an English company which performed that service in a very different manner from the present.

Brazil has no lack of swift steamship communication with England, for England has taken good care that there should be no such lack. The three principal English lines employ not less than fifty-five steamships in the trade between the two countries. Some of these lines have their steamships return to England by way of the United States, thus intending to break down any competition that may be attempted by our merchants. The French companies keep nineteen steamships plying between France and Brazil, and a German line employs fifteen ships in the Brazilian trade. As against this showing for Europe, what facilities of communication with Brazil has the United States? One line, established in 1878, and since maintained in the face of government discouragement and the fierce opposition of the English lines. This new line, employing the finest ships and making one trip per month each way, has given to American merchants their first direct and quick communication with the South American markets. Is it any wonder, in view of these facts, that we have had little trade with Brazil? Let us see how this state of affairs has been brought about.

An Example to be Imitated.

It has for many years been England's policy to encourage the opening of new markets wherever there was chance to get possession of trade, and especially where the circumstances were such that the dependency upon her for supplies could be made continual. Once in possession she knew it would be comparatively easy to keep the business. The first step in her policy was to establish swift steamship lines between the two countries, thus bringing their business men together, and giving the merchants of the new market quicker mail communication, and better, cheaper, and faster passenger transportation to England than to any other country. This was imperatively necessary to secure the desired trade, and it was in every case successful. The next step was to exert the needed influence to induce the foreign government to meet her half way.in the liberal policy of encouragement. The figures given above show how this plan has worked to England's interest in her relations with Brazil. Besides securing the means of quick orders, collections, and delivery of goods, she gets for her merchants more than a million dollars a year from Brazil, to help them drive the ships of other nations out of the trade. We shall see a little further on how they attempt to accomplish that end. Let us now look at a different policy.

OUR COMMUNICATION VIA ENGLAND.

It seems incredible that the United States should have been content to communicate with Brazil by way of England and entirely through English companies, but so it was. Even as late·as our Centennial there was no direct line from this country to Brazil ; and when that enlightened Emperor, Dom Pedro, wished to visit our shores and see for himself our national progress, he was obliged to go first to England, thence hither. The mail contract now held by an American citizen was then in possession of an English line, though in a different form. In the old form, Brazil paid the English line $100,000 a year to carry the Brazilian mail from Rio to New York, making no stops at any other South American ports, in slow ships of 1,000 tons, with no decent cabin accommodations. Let us see how skillfully was laid a great pipe, so to speak, through which under this contract all the gold and profits of the South American trade should flow into England. The English ships started from Liverpool with a cargo of English manufactures for Brazil. At Rio the freight was collected, and a cargo of coffee taken on for the United States. That cargo was delivered in New York, and the freight collected there in gold. Then the ship loaded up with a cargo of breadstuffs for Liverpool, and sailed back to the home port, there collecting the third freight, and completing the profitable triangular trip ; while at the same time bringing home with it the gold drained both from Brazil and the United States for freights. We were left without direct return communication. Our mails had to go to Brazil via England, and to pay the Brazilian merchants for their coffee we even had to send gold to them through the English bankers, paying them the commission for exchange. By this means England got not only the freight moneys and all the profits, but also the price of the cargo ; for her bankers first took their commission out of the gold we sent to Brazil, and her merchants subsequently received the balance of it for English manufactures. It will readily be seen that we were thus effectively prevented from getting our products into the Brazilian market. While carryiug out this shrewd plan the English line, moreover, was drawing mail pay from Brazil to aid it in driving away competitors.

THE AMERICAN LINE TO BRAZIL.

This was the singular condition of affairs when the Emperor of Brazil came to this country. He readily appreciated the situation and the natural advantages for trade with his country which we possessed. To establish closer commercial relations between the two countries, the Emperor agreed to transfer the contract, which was soon to expire, from the English line to an American citizen, but on very different terms. Under the new contract the ships, instead of being 1,000 tons, slow, and without cabin accommodations, were to be of 3,500 tons, of great speed, with first-class cabin accommodations for 100 passengers, and equal in every respect to the finest ships entering Brazil from any European nation; so that the facilities for trade with the United States would be equal, so far as they went, to the best. It was expected by the Emperor and the Brazilian government at the time the contract was made, as well as by myself, that a line of steamers of this character could be run and carry the mails for the same compensation both ways which the English line received; and that, if American capital could be found to put American ships on the ocean, to develop a new and desirable trade, and to perform a needed service without asking more pay for it than England, France, and Germany were giving to their mail lines, the United States would certainly be only too glad to meet Brazil half way in the establishment of direct and reliable communication. What was the result?

After I had thoroughly examined into the subject and the possibilities of successful enterprise, I went to the large capitalists of New York and Boston—the men of all men in this country who would be likely to invest in such a movement—and presented the matter to them. I discussed the question with them in every point, showed them the great value an American line would be to our trade, and spared no effort to induce them to join in establishing such a steamship line to Brazil as they were ready enough to admit the country needed and must have. But I could not raise a dollar, owing to the fact that they saw the difficulties of establishing business in a new country—no matter how great our natural advantages—and particularly in a country controlled as to trade by an opposition so formidable as that of Great Britain. Even after the encouragement given by the Brazilian government they would not invest, nor was a dollar secured until the ships were built and put on the route.

MAKING THE START ALONE.

The contract with Brazil was made, however, and the line was established, in the expectation just alluded to. How the terms of the contract were

complied with may be seen in the following extract from the recent report
of Mr. Adamson, United States Consul-General at Rio de Janeiro :

"The United States and Brazil Mail Steamship Line is the only one plying regularly
both ways between the United States and Brazil. The first trip was made in May, 1876,
and since that time this line has given us one steamer per month each way, making the
voyages with great regularity in from twenty-one to twenty-two days generally. The
three fine ships of this line are of 3,548, 3,532 and 2,764 tons burden respectively. They are
unequalled by any ships entering this port in convenience, comfort, and in their appoint-
ments in general."

The additional fact should be mentioned that, while the little English
line only brought the mails from Rio, the American line (which has never
missed a trip during its existence) has brought mails from Rio, Pernambuco,
Bahia, Para, and St. Thomas.

A POLICY OF OPPOSITION.

Yet in starting this enterprise, without any assistance whatever, I never
expected aid from the government, nor asked for anything but reasonable
compensation for carrying the mails, to build up a business which would
benefit greatly the whole country—a compensation such as the European
governments which have trade with Brazil pay to their mail lines, as stated.
The Brazilian government did its half, and transferred the $100,000 con-
tract to the American line. How did our government meet this offer to
establish closer commercial relations with a people from whom we buy six
times as much as we sell to them in return? Immediately the howl of
subsidy and monopoly was raised, and our Congress saw fit to refuse the
return contract. Could any policy have been pursued that would help
England more to hold her position and keep us from rivalry?

When this enterprise was open to our capitalists, as I have shown, and
is still open to them, but they could not be induced to invest, where in the
name of common sense is the monopoly? As to subsidy, meaning by that
the fair pay due for service rendered, from the government to this struggling line, which is certainly more important for the interests of the great
mass of producers than it can be for those of any individual, it is not only an
unjust but an exceedingly short-sighted policy that withholds such encouragement. As reward for maintaining such a line I am met, too, with such
arguments—if they are worthy the name—as this, taken from the Evening
Post of recent date. Answering a correspondent who claims that to make
competition fair for American steamship lines foreign subsidized steamers
ought not to be admitted on equal terms to our ports, the editors say :

"Our correspondent is savage. Foreign steamship lines have never 'prevented American enterprise from building steamships.' Our infernal tariff and Navigation Laws have
prevented us from either building or buying steamships. Having thus driven ourselves
from the ocean, 'South Street' wants us to drive all the foreign commerce from this port.
If this barbarous plan could be accomplished the merchants and clerks of South street
would probably have a permanent vacation from business."

This is the spirit and language in which this subject is treated by an old
and respectable newspaper. How true the blame cast upon the "infernal

tariff and Navigation Laws" is I have shown you pretty fully hitherto. It is most strange that a paper established in New York for eighty years has not yet found out what is for the real interests of that city, but is wholly wild on the hobby of free trade.

For their part the English lines were not quiet, but alive and fierce in their attempts to crush out the new line. No sooner had Brazil transferred this one contract for carriage of her mails to our line, than the merchants of Liverpool formed combinations to defeat that line in its efforts to secure trade. They regarded it as an intruder, and saw that unless they could crush it they were threatened with a serious rivalry, particularly if the United States met Brazil as England had years before. The English line offered to carry the Brazilian mails to the United States for nothing. That failing, the cutting of freight rates was begun. Compare the present rates with what they were two years ago. When the American line started the freight on coffee was seventy cents a bag, or $12 a ton for carriage of 5,200 miles. In the first year of the new line, it was reduced to sixty cents a bag, or $10.25 per ton. When our government refused to pay its own line for mail service, the English opposition were jubilant, and concentrated their efforts to prevent the Brazilian Chambers from appropriating the money for its contract except on such conditions as could not be complied with by the line. Pains were taken to have the more violent speeches made in our Congress against this line translated into Spanish and laid upon the desks of the Chambers when the appropriation was under discussion. It was charged that the Americans did not want the trade of Brazil, and the Brazilian merchant might well believe it when he read of the manner in which the proposition to open facilities for trade was met by some of our representatives and newspapers. The result of all this was that changes were made rendering the contract worthless to the American line, which the English thought would then be withdrawn, and serve as another warning, similar to that in the railroad instance given above, to all other American capitalists against attempting to interfere with English interests in Brazil. But to the honor of the Brazilian government be it said that the law affecting the contract was decided by the Supreme Court of Brazil in favor of sustaining the contract without change until the Chambers meet again to confirm their action. The next move was begun this year, and is in the shape of another reduction of freights on coffee. Now the rate is only thirty cents a bag, or $5.25 per ton for 5,200 miles—a ton of coffee 1,000 miles for $1, or less than it costs to discharge the same in New York and transport it to any point within fifty miles. And this is what the American line has to compete with, in addition to the discouragement and abuse at home. But this is not all.

Unjust Discrimination.

While our government refuses to pay this line fair compensation for mail service, it compels that service for sea-postage merely. Under our laws an

American citizen who runs an American ship to a foreign country must
carry the United States mails for the postage ; but should that citizen
place this same ship on any American coast, river or lake route, the government would be by law compelled to pay a liberal compensation for the mails
carried by the ship. Where is the consistency? Would the government
for a moment think of compelling the railways to carry the mails without
compensation? What is it that makes such a distinction between the ship
and any other carrier?

American merchants are willing to furnish the capital and build ships at
home, and thus give to our mechanics the benefit arising from this labor ;
they are willing to pay taxes at home and keep the capital in the country ;
but they are not willing to invest this capital and then be compelled to give
gratuitous service to the country, and ruin themselves. It is an arbitrary
law which compels the mails of the United States going to a foreign country in an American ship to be carried free, when if the same ship was placed
on an American coast route the payment for mail transportation would be
liberal. English capitalists are willing to carry the American mails to
countries where Americans cannot interfere with their trade, but they have
never yet proposed to give us a service to Brazil, for the reason that it
would injure English interests at home.

WHAT THE NEW LINE HAS DONE.

The American line has given the American merchants for the first time
direct, rapid, first-class means of communication with Brazil and the other
South American markets. It has given, therefore, what they never had before, the chance to put their goods into those markets. Passengers from the
United States to Brazil now go from New York to Rio direct, where they
used to go to England first, making a long and expensive trip, useless for
purposes of extending trade, while all the money went to English ships.
It was the same route, via Liverpool, for passengers from Rio to New York,
with pay also to the English line. Now the passenger money from both
ways comes to this country. The freight money which was formerly so
skillfully gathered by the English company both in Rio and New York, is
now collected by the American line, so far as it is able to perform the service. For instance, the new line takes a cargo of our products from New
York to Rio; there collects the freight money, takes on a cargo of coffee,
and brings it to New York, collecting the freight in that city. All this
money goes into home circulation. Is there no value to the country in this,
as compared with the plan of the English line? Will it be of no loss to our
people if this line is withdrawn, as it must be if government refuses to adopt
a more far-sighted and liberal policy? Mr. Adamson said, when it was
feared that this line must be taken off because of the adverse action of the
Brazilian Chambers, founded on the adverse action of our own Congress—
that this would be an "event very much to be deprecated, as it would

inflict a fatal blow on the efforts recently made to increase our export trade to Brazil. Unless dealers can depend on getting goods out within a definite time, they will not buy of us while the certainty exists that orders sent to England can be filled and shipped within the same week." He added this most significant paragraph, which I urge upon your attention:

"Just as it became probable that our American line must be withdrawn, an agent of our Canadian neighbors appeared on the scene and announced that the Dominion government had agreed to pay $50,000 per annum toward a line between Halifax and Brazil. He has now gone home, having obtained a contract from this government, and we are promised the first ship of that line within four months. The Dominion government proposes to follow this effort for Brazilian trade by opening here in June next an exhibition of Canadian wares and products, which will not be a private enterprise, but entirely under the control of the government. Meanwhile various projects for an American Exhibition have been announced, but the date of opening does not appear to have been decided on. It is to be hoped that the people of the United States will not allow themselves to be outdone in enterprise by their Canadian neighbors."

REWARDS OF ENTERPRISE.

I have thus given in brief the history of the new American line to Brazil. In view of the easy availability to us of the great South American markets and of our need to find such outlets for our surplus products; in view of the liberal course pursued by the Brazilian government (which certainly has little interest in the extension of trade between the two countries compared with what ours ought to be, since Brazil already sells her surplus to us, and has every facility offered her to buy what she needs from England) in reference to the establishment of a direct steamship line to the United States, was it unreasonable for me to believe that our government would meet Brazil half way in giving the encouragement necessary? I have shown you how that expectation was realized, and also the opposition and sort of competition under which the line has struggled. In the face of these facts, and of the utter failure of the efforts to get our capitalists to invest in such a risky business, perhaps it will not be so frequently charged that it is a sufficiently profitable thing to start such an enterprise, and that to give reasonable pay for mail service actually and excellently performed would be to spend the people's money unnecessarily and extravagantly. The probability is that the opposition will be successful, and that the only American line to Brazil will be withdrawn. It certainly must be unless our government changes its policy. Then we can return to the old order of things, and there will be no more outcry about subsidy or monopoly. But what will be the effect of such a step on the interests of this country? Was ever so small and false economy practised by so great a people? I cannot believe that we shall permit Canada to take possession of that trade with South America which is now within our grasp. But the English interest is powerful, and has accomplished great things, as we have seen, while our people have apparently been charmed by it into a condition of inactivity favorable to its, but fatal to their own, interests.

Something that Requires Time.

It is very true that the trade between the United States and Brazil during the last two years has not increased so much as might be desired ; but it would hardly be just to use that as an argument against the support of the American line. It is to be considered that Brazil in these two years has suffered from a great famine, so that her people were unable to purchase as largely as formerly—their currency, also, being at a discount of 25 per cent. Then the depressed condition of the English manufactures enabled the English merchants to place their goods in the Brazilian market cheaper than formerly, while the advance in prices in this country prevented us from shipping our goods there. Above these reasons, all business men know the great difficulty of entering a new market in competition with the English, French and German merchants who already had possession ; and they know it is not fair to expect that a large trade will be built up in a month or a year. Nevertheless, despite the unusually bitter opposition from abroad and the still more discouraging opposition at home which this new line has had to endure, it can be said that in the eighteen months of its existence more merchants from Brazil and the South American coast have come to this country than came here during the whole ten years previous ; while more of our people also visited the South American ports. The statistics will show, further, that we have sent to Brazil in these eighteen months four times as much of the products of our factories as formerly.

Important to the Producers of the Mississippi Valley.

The fact is worthy of note that there is a specially good market in Brazil for certain brands of flour, more particularly those made in Richmond and St. Louis. The present line has in eighteen months carried over 180,000 barrels of flour. Fifty thousand barrels of this came from St. Louis to New York. This was a great drawback both in time and cost ; and though the freight from New York to Rio was only from eighty cents to $1, when the cost of transportation from St. Louis to New York was added, the price of the flour sold at reasonable profit in Brazil was so great that the grain growers of the Argentine Republic could undersell the American flour shippers. This exactly served the English interest and fitted into England's policy of getting all the gold ; for she has superior steam mail communication with Buenos Ayres, the same as with Rio, and gets the trade of the Argentine Republic by the same attention to her foreign interests which secured her that of Brazil. Our commercial relations, on the other hand, are equally as unsatisfactory with that Republic as with Brazil, though on a smaller scale. For instance, as the table shows, we buy from the Argentine Republic $5,834,709 worth of products where we sell but $1,351,294 in return ; and they send the balance of gold which we pay them, to the European nations for their manufactures. The only difference to England in the case of the flour was, that we bought coffee from Brazil and paid her in gold instead of flour. She paid the gold to the Argentine Republic in exchange for flour; and the Argentine merchants sent it to England in exchange for manufactures. As hinted in the quotation already given from the Liverpool Circular, it is in part upon the gold received in exchange from the South American markets that England depends. Now, I ask again if it is not to our great interest to keep that gold in this country by sending our flour and other products, instead of our gold, to those markets in which we are the largest purchasers?

Suppose a semi-monthly line to Brazil was running from New Orleans to Rio de Janeiro, giving a direct outlet to the vast products of the Mississippi Valley. The producers of that most important section of our country would thus save the time and freight from St. Louis to New York, and could put their bread and meat in South America in such time and at such price as would control those markets. In addition they could bring back in exchange the coffee and other products demanded from South America, and save again in the cost of transportation from New York to the consumers of the Mississippi Valley.

Not Pleasant to Reflect Upon.

To sum up this matter, it does seem almost too singular to be true that, with two so great countries as these, Brazil, the second American nation, and the United States, the first, engaged in trade with each other, the latter buying each year one-half the surplus products of the former, amounting to between $40,000,000 and $50,000,000, and paying for them principally in gold, not one dollar of that vast sum can be settled for in the metropolis of the United States, but must be settled for in London, the American purchaser and producer both paying commission to an English banker, paying the freight to an English ship owner, paying the gold for the coffee to a Brazilian merchant in order that he may spend it in the English market. It seems stranger still that no longer than two years ago the American merchant who desired to visit Brazil must first go to London and from London to Brazil, or send his letter by the same roundabout way; and that the Brazilian merchant wishing to visit the United States must proceed in the same expensive and time-consuming way—which resulted in his buying what goods he needed while in London and shipping them home before he came on to our shores. Yet this was the condition of affairs before the American line was started; and it will be their condition again if that line has to be withdrawn by reason of the treatment it has received at the hands of that government whose name it bears.

An Interest too Great to be Individual.

For what I have said on the subject of our carrying trade, I may be criticised as being a ship builder and owner, directly interested in the revival of our shipping. But it is to be considered that these questions of ocean carrying and opening new markets will continue to be of great importance to this country when all the men of the present generation who are interested in their solution shall have passed away. Moreover, what is a nation but an aggregation of individuals ; and what progress was ever achieved that was not the consequence of individual effort? The prosperity of the nation is made up of the industrial results of its members, and while the failure of one, two, or three may not be felt, the nation's growth and greatness are nevertheless dependent upon the success of its individuals. All measures of national interest must benefit individuals, but surely such measures are not for that reason to be abandoned. Nor is it rational to suppose that, if a policy is adopted that will induce capitalists to invest their money in ships and ship building, the business will be kept to any one man or set of men in a great nation like ours. There will be no question of monopoly in ship building when that industry is found to afford a chance for profitable investment. There is room and freedom for all, as I have repeatedly said, both in that and in any other of our great industrial interests.

The Same Policy for All.

The necessity to commerce of rapid communication on the land has been nowhere appreciated more than in this country. We have spared no efforts to reduce distance there to short time. If our carrying be on the sea instead, is there less necessity for swift transportation? Why shall we not furnish with equal care the means for the rapid reaching of these outside markets? The conditions and requirements of trade are not changed because the separating distance is covered by water. We must not think, because the road to market lies on the water, that we can depend upon the old wooden sailing ship while foreign nations are supplied with iron steamships that could take out a cargo and return while we were creeping along, as wind and tide might allow, toward the outbound port. That would be like pitting the old-time mule team and wagon against the freight train, and the stage coach against the lightning express. When our capitalists and ship owners simply ask that to capital invested in the foreign carrying trade the same policy with respect to mail service shall be applied which is applied to capital invested in our coastwise and land carrying lines, will any public man permit himself to be driven from doing his duty to the country by the outcries of subsidy, monopoly, and free ships, raised by those who are interested in limiting, not extending our carrying trade?

Through the policy of being middleman for the nations, England has succeeded in drawing the wealth of other countries to herself and in retaining that wealth at home, thus making herself the banking house of the world, though she does not possess the natural advantages which we do. We have a class of surplus products which the other nations must buy from us. If we pursue a course that will draw and keep the wealth of the world here, we shall be able to furnish cheap money to the West and South, and their producers can then undersell all competitors in the markets of the world. There would be no difficulty in that case in disposing of our surplus, however much it increases.

It ought to be said in this connection, as a warning for future action, that in what little attempt has been made by the United States to meet England's policy of encouragement, we never guarded the interest of the government by requiring first-class ships of great speed, able to compete with the best ships of other nations. As a general thing, most unfortunately, the contracts that were made were given to speculators who got a lot of old worn-out steamers, put them on the routes, and made a stock speculation out of it. When the contract expired the trade on these routes was naturally in no better condition than when the lines started. Yet this failure of enterprises which were never undertaken in a way that made success possible, was used as an argument why nothing more in that direction should be done ; whereas it was known by every practical man that the lines were frauds from the start, and that no benefit to the country was to come out of them. It is blind logic to say that because a thing was a failure when done in a wrong way, no effort should be made to do it in a right way ; particularly blind when it is considered that so many interests of the greatest country in the world are involved in the doing of that very thing right and successfully.

The Free Ship Man's View of Labor.

To return for a moment to the table on page 67 comparing the wages paid in an English and American ship yard, it is the favorite argument of the advocate of free ships and free trade, that the English workman can live as much cheaper in proportion as his wages are lower, yet with equal comfort. Investigation of the figures will show whether that view is reasonable. The range of prices for work on the ship in Great Britain is from $8.22, the highest (and this for skilled labor), to $1.25, for common labor, per week. In America the range for the corresponding classes of work is from $19.80 to $3.30, or two and one-half times more. Will the free trader claim that there is as much difference as that between the cost of living in England and America ? Or that a laborer can live in England and support his family on 21 cents a day ? The average English rate, moreover, for the thirty-six classes of workmen is $5.35 ; the American rate $11.27. The rates are too unequal for the free trader to overcome by a mere statement. Besides, the English workingman lives on American bread, and on American meat, too, when he can afford meat. If the free trader's claim that the English laborer can live on his wages is correct, it simply means that he has no such home as the American workingman has—which is the fact. As it is also the fact that the American workingman is the only one in the world who has a home in the true sense of the word, and the same rights as all other men ; and that the foreign workingman only remains where he is largely because he cannot get the means to emigrate.

This is the only land where the workingman can look forward to establishing and owning for himself a home where he may spend his old age. It is the aim of our policy as a nation to give the workingman this chance. It is nonsense to say he can have such chance on $1.25 a week, or even on $8. How can a man clothe and feed himself and family, pay rent, and have anything left for the rainy day out of such a sum as that? Especially when it is considered that his days of "no work" and of sickness must be deducted. This claim of the free-trader is absurd. The differences are too great.

WISE WORDS.

In immediate connection with the subject of our shipping, what were the views of our early statesmen, who shaped the policy which made us, while yet in infancy, the second carrying nation of the world? In regard to the necessity of possessing ample means of national defence on the ocean, Washington, in his message to the third Congress, said: "There is a rank due to the United States among nations which will be withheld, if not absolutely lost, by the reputation of weakness. If we desire to avoid insult we must be able to repel it; if we desire to secure peace—one of the most powerful instruments of our prosperity—it must be known at all times that we are ready for war."

PROPHETIC UTTERANCE.

In his Report on Commerce, Jefferson said:

"Our navigation involves still higher considerations. As a branch of industry it is valuable; but as a resource of defense essential. The position and circumstances of the United States leave them nothing to fear from their landboard, and nothing to desire beyond their present rights. But on the seaboard they are open to injury, and they have there too a commerce which must be protected. This can only be done by possessing a respectable body of citizen seamen, and of artists and establishments in readiness for ship building. If particular nations grasp at undue shares (of our commerce), and more especially if they seize on the means of the United States to convert them into aliment for their own strength and withdraw them entirely from the support of those to whom they belong, defensive and protecting measures become necessary on the part of the nation whose marine resources are thus invaded, or it will be disarmed of its defense, its productions will be at the mercy of the nation which has possessed itself exclusively of the means of carrying them, and its politics may be influenced by those who command its commerce. The carriage of our own commodities, if once established in another channel, cannot be resumed in the moment we desire. If we lose the seamen and artists whom it now occupies we lose the present means of marine defense, and time will be requisite to raise up others, when disgrace or losses shall bring home to our feelings the evils of having abandoned them."

These words seem well-nigh prophetic in the light of events, except that the danger which enabled other nations to seize undue shares of our commerce did come from the landboard, through our civil dissensions; but had we been prepared to meet the outward pressure brought to bear by England, we could not have been reduced to our present condition.

A Strong Reason.

Mr. Madison said, in reply to the free trade arguments of his opponents in 1794, that he was friendly to a free intercourse with all nations:

"But to this rule there might be exceptions. The Navigation Act of Great Britain had secured to her eleven-twelfths of the shipping and seamen employed in her trade. Here was a great gain from a departure of the rule. Another exception to the advantages of free trade is found in the case of two countries in such relation to each other, that the one, by duties on the manufactures of the other, might not only invigorate its own, but draw from the other the workmen themselves. To allow trade to regulate itself is, as our own experience has taught us, to allow one nation to regulate it for another."

Further he said that "we should not be in a state of commercial dependence upon a single nation for necessary articles of consumption or of defense in time of war." The views and advice of these staunch and able statesmen are applicable to our affairs to-day. Were the founders of our government living, would there be any question with them as to what policy should now be pursued? Would they consent that we should remain inactive while foreign nations are steadily increasing their control over trade which we ought to successfully compete for and secure, and while our carrying interests are being still further jeopardized by delay?

XVI.

THE AWAKENING.

The time has come, happily for our country, when the vast importance of these questions will no longer permit them to be overlooked. When the people at large begin to understand more fully our position and policy in relation to the carrying trade on the ocean, in the light of England's position and policy, and see what she has accomplished by statesmanlike legislation and careful fostering of her shipping interest; when they realize that we have superior natural resources to enable us to own home-built ships, whether iron or wooden, and that to build the ships at home means to give ninety per cent. of the large capital invested to American labor, and thus to distribute that capital among our own people and maintain a profitable home market; when they consider that at least one-half of the $70,000,000 or more which we now pay annually in freight and passenger money to foreign carrying lines (owned by foreign capital, supporting foreign labor, and paying taxes to maintain a foreign government), might be paid and most certainly ought to be paid to American lines (owned by

American capital, supporting American labor, and paying taxes to maintain our own government), I am convinced that they will never consent that this immense sum of money shall continue to be sent from this country to support foreign nations, but will demand an entirely different policy, framed in our own interests. They will no longer be satisfied with inactivity, but will require such vigorous action as shall enable our merchants and builders to meet England on the ocean. When, furthermore, our people become more thoroughly informed as to the foreign interests which lie behind the free ship movement in this country, no matter how honest it may appear on the surface, nor how many unquestionably honest men are mistakenly among its advocates, they will no more have patience with this outside effort to dwarf our development as a great manufacturing and industrial nation, and to keep us from the place which belongs to us as a carrying power on the seas.

Drawing Conclusions.

From what has been said, these conclusions may be drawn : England's Navigation Laws established her power on the ocean. She kept them in force 200 years, and under them became the first carrying nation of the world. In 1850, when she had secured for her people the power to build iron ships cheaper than they could be built elsewhere, and there was no demand for such ships as we could supply at competitive or desirable rates, her Navigation Laws were repealed.

Under our Navigation Laws in earlier days we astonished the world by our progress and triumphs on the seas. We became a carrying power second only to England. I have shown that the loss of our position as carriers can in no way whatsoever be attributed to the Navigation Laws. On the contrary, but for them the promising start we have made in iron ship building since the war would have been impossible. We owe the present depressed condition of our carrying trade to our failure as a government to meet England's policy of encouraging the building of iron steamships and the establishment of fast mail lines; to our long and costly civil war; and to the indifference of our government in regard to this great carrying interest. No fault can at any point be laid upon the Navigation Laws. To repeal them now would be to stop us where we are as ship builders, and leave us, in company with the other nations of the world, dependent upon one country—England—for our ships and tools. When our people receive the same encouragement from our government that the English builders and merchants have had from theirs, and when a liberal policy is carried out in relation to our carrying trade, we shall not need to wait so long as England did before we may profitably repeal them. It would be fatal to our shipping and industrial interests to do it now.

I have already stated that there are eight millions of men in arms in Europe, supplemented by powerful navies ready for action. To support these vast standing armies last year cost Russia $180,000,000, England $160,000,000, France $135,000,000, and Germany $105,000,000—a grand total of $800,000,000, spent in time of peace to maintain the facilities for war. This does not speak well for peace in the future. Who can foresee what outbreak may take place in Europe at any moment? While there is no reason why we should be drawn into a foreign war, there is every reason to believe that our ocean road to the world's markets would be interfered with by such a war. Is there any excuse for our running the risk we do in depending upon foreign carriers to deliver our surplus products in those markets? If it is possible for us to place this important business within our own control, so that either in war or peace we can be self-reliant, ought we not to set about it immediately? And who should be more interested in that enterprise—the capitalists who have their money in hand ready for investment in whatever line they may select, and are independent whether or not we build ships to do our carrying, or the great masses of the producers of the country, who are dependent upon the means of transportation to market? Who should care most whether that transportation is under our flag, or controlled by a foreign nation and thereby subject to all kinds of European war risks?

There can be no question, I believe, that the people of this country generally are eagerly desirous that something should be done at once to revive the carrying interest. But as yet they are divided as to what is the best thing to do. They have been divided largely by being told that this end was to be accomplished only by repealing our Navigation and other needed protective laws. For so sweeping and far-reaching changes as these the more thoughtful and informed are by no means prepared, and very fortunately for the country's welfare cannot be brought to look with the necessary favor upon them. I hope I have been able to make it plain to every reader of this pamphlet that the desired revival of our carrying trade would in no wise result from such changes, but would be absolutely prevented by them; and that the real remedy lies in such just legislation as shall remove all obstacles to equal competition with our rivals, and enable us to meet them any and everywhere; legislation that will give us like advantages with people who are sustained in their national enterprises by their governments.

NATIONAL LEGISLATION.

It will be a hopeful day when our people fully recognize the fact that the carrying trade is a national interest, to be governed by national laws. As it is now, much of the trouble in connection with the growth of that trade in American bottoms is caused by local taxation, one thing in New York, another in Baltimore, and so on, under the State laws. This foreign trade ought to be under a comprehensive national law, and its interests zealously cared for by national legislation. For though the ships may for reason of cost or convenience enter at certain ports, the precious cargoes they bear are gathered from and dispersed among all sections, and in the economical carrying and safe delivery of those cargoes all sections are alike interested. It may not make any difference to the farmer or shipper in Ohio what you tax a house and lot in New York, but it does make a difference to him how much you tax, under your local laws and for your local benefit, the ship which carries his product and that of every other shipper in every State of the Union; and that tax makes a much greater difference still to the owner of the ship, since it so discriminates against him and so burdens his property that he cannot profitably compete with his foreign rival in carrying. It acts, indeed, as a prohibitory law, so that nobody derives any benefit from it, while it injures the interests of the whole country. How can there be competition while there is imposed upon the American ship in the American port a local tax from ten to twenty times greater than the tax laid upon the foreign ship in its home port; and while that foreign ship can freely enter our ports, being beyond the reach of local laws? Can there be any plainer duty than that every local difficulty in the way of our regaining place as a carrying power should at once be removed?

In the legislative treatment of the shipping question our statesmen labor under the disadvantage of having to meet opposition where aid and hearty co-operation should be met instead. Legislating from a sectional standpoint is one of the weakest features in our system of government. England's carrying strength is due in great measure to the fact that this interest has been dealt with in her Parliament purely as a national question. When she was trying to build up this interest there was no division into England or Scotland, London or Liverpool—it was all Britain. When legislation is discussed in our Congress, there is at once competition and rivalry between the various sections, each wanting its own particular needs attended to first; and frequently strong antagonism is manifested on the part of the interior districts to any measures proposed for the sea-board, as if the whole country were not involved in the development and growth of every part and interest. What section is more concerned than the agricultural West and South in having a good and ready market for their products, and a swift and sure means to reach that market? Or in having one of our great cities (instead of London) the world's banking house, thus securing cheap money for the American farmer, as well as for the American manufacturer and business man of whatever class? This great country of ours is like the body, every vein and artery of which leads to a common centre. Every section of our land is one of the great arteries through which flows the life-blood of the nation, and upon the health and vigor of every part depend the health and vigor of the whole. There can be no division of interests without danger and detriment to all, any more than there can be prosperity and benefit to one part without the sharing of its effects by all the rest.

An American Policy.

The carrying trade is eminently and intensely a national interest, and in acting with regard to its rebuilding and recovery we should legislate, not for a section, but for America, and leave it for the capitalist to be judge as to where he can most profitably invest. It is of the greatest importance that no false step should be taken. If our carrying trade is to be restored, it should be restored on a solid basis, which will secure it to us so surely that no circumstance of war, and no foreign power nor all foreign powers combined, could again reduce it to its present low condition ; or, if that is not possible, at least to put and maintain us in position to rebuild immediately within ourselves.

I believe it will be conceded by all of you that these questions, How to regain our rightful place on the ocean, and to open up new markets for our surplus products—are vital issues, which demand from our statesmen careful, calm, thorough consideration, free alike from sectional bias and the influence of outcries about monopoly and subsidy. Heretofore measures have been introduced into Congress to secure special contracts for mail lines on the ocean similar to the contracts made by Great Britain and other nations under their policy of encouragement, a policy which they have long found profitable and still pursue. In foreign countries these measures receive due consideration and their benefits are recognized. With us they have not found favor, but without consideration have been greeted by the howl of subsidy, lobbying, and monopoly. Therefore, when legislation is reached on these great questions, let there be no chance given for the renewal of that howl. Let us be even more broad and liberal than England, France, or Germany, for their legislation has been in the interest of responsible private enterprises. Though by such a policy they have been most successful, let us not follow their example in this respect, but make our national policy so free and fair that the enemies of our carrying trade, who are seeking every means to prevent its revival, can find no ground or excuse for opposition. Let our carrying policy be one that will invite the widest competition not only from the capitalists of our own great country, but even from the capitalists of foreign lands who wish to invest under our laws.

Very Respectfully,

JOHN ROACH.

www.ingramcontent.com/pod-product-compliance
Lightning Source LLC
Chambersburg PA
CBHW032247080426
42735CB00008B/1036